THE LAST
RAIN FORESTS

THE LAST
RAIN FORESTS

A World Conservation Atlas

Edited by Mark Collins

Foreword by David Attenborough

New York
OXFORD UNIVERSITY PRESS
1990

Preface

"Delight ... is a weak term to express the feelings of a naturalist, who for the first time, has wandered by himself in a Brazilian forest." Charles Darwin wrote these words in February 1832. More than 90 percent of the luxuriant coastal forest that he explored has gone. Recent estimates suggest, moreover, that nearly two percent of the world's remaining rain forest is lost every year. Unless this trend is halted there will be little left in 50 years' time.

IUCN recognizes the urgency of development and the need to use nature's renewable resource systems to satisfy human requirements and eliminate the appalling poverty that afflicts so many people today. But such development must be *sustainable*, using the income from nature without eroding its capital of soil, water and fertility.

To many tropical countries, forests are an important constituent of national wealth. They protect soil fertility, regulate and purify water flow, and yield valuable timber and other products; used sustainably they are an economic and environmental asset for tomorrow as well as today. Used destructively, for short-term gain, they are all too likely to be replaced by degraded land and impoverished people. Developing countries cannot afford such waste.

Development must be based on a sound understanding of environmental systems. This atlas sets out our knowledge of the forests that remain. It also explains the complex issues that must be considered when making decisions about development in rain forest areas. Neither this book nor IUCN argues that rain forest should be sacrosanct everywhere – in some areas, conversion to intensive agriculture or agro-forestry may be the best course to pursue. But in much of the forested region, a balance between protection and skilful sustainable use is needed.

If this balance is to be achieved, the industrialized countries that buy tropical forest products must support and cooperate with producer countries. They must look again at how the burden of debt forces tropical countries to clear their forests in order to secure immediate income. They must help developing tropical countries to implement sustainable management and appropriate pricing systems, so that rain forests remain to enrich the lives of future generations. And those making plans for the future must be sensitive to the possible global impact – climate change and rise in sea level – that could render current conservation plans inapplicable in the future.

The attention paid to rain forests by the world's media is an encouraging sign of the growing general concern for environmental conservation. But some campaigns have failed to see the issues from the point of view of the people who inhabit the developing countries of the tropics. Saving rain forests requires a full understanding of their complex ecology, but depends even more on finding solutions to human social problems – alleviating poverty, easing debt, and above all creating conditions under which human population growth will slow, and people achieve an enduring harmony with nature. This book provides an insight into these vital issues.

by Martin W. Holdgate,
Director General of IUCN

General Editor
  Dr Mark Collins

Consultant Editor
  Jeffrey Sayer

Cartographic Advisor
  Mike Adams

Research Assistant
  Clare Billington

Contributors
  Deni Bown
  Robert Burton
  Dr Mark Collins
  Dr Roger Cox
  Dr Linda Gamlin
  Dr Caroline Harcourt
  Dr Nick Middleton
  Dr Alison Rosser
  Andi Spicer
  Charles Tyler
  Dr Timothy Whitmore

ISBN 0-19-520836-6

Printing (last digit): 9 8 7 6 5 4 3 2

Published in the United States of America by Oxford University Press, Inc., 200 Madison Avenue, New York, N.Y. 10016

Oxford is a registered trademark of Oxford University Press

Edited and designed by Mitchell Beazley International Limited, Michelin House, 81 Fulham Road, London, SW3 6RB

Executive Editor   Robin Rees
Senior Art Editor   Paul Wilkinson
Editor   Simon Ryder
Assistant Editor   Mike Darton
Map Editor   Frances Cockayne
Picture Research   Andi Spicer
Production   Ted Timberlake

Maps by Lovell Johns Limited, Oxford
Indexed by Annette Musker

© Mitchell Beazley International Ltd 1990
All Rights Reserved

Typeset in Century Schoolbook by Servis Filmsetting Limited, Manchester
Reproduction by Colourscan, Singapore
Printed and bound in Hong Kong

Library of Congress Cataloging-in-Publication Data
The Last Rain Forests: a world conservation atlas / edited by
  Mark Collins; foreword by David Attenborough.
      p.   cm.
  Includes bibliographical references and index.
  ISBN 0-19-520836-6
  1. Rain forest ecology. 2. Rain forests. 3. Rain forests–maps.
  4. Rain forest conservation. 5. Rain forest conservation–maps.
QH541.5.R2 L38   1990
574.5′2642′0913–dc 20   90-226854
                                                              CIP

Measurements – Both metric and imperial measurements are given throughout.

Billions – Because of the differing usage between the United Kingdom and the United States, billions here correspond to thousand millions: £15,000 million = £15 billion.

Brazilians now talk of three seasons: the rainy season, the dry season and the *queimadas*, or burnings.

Contents

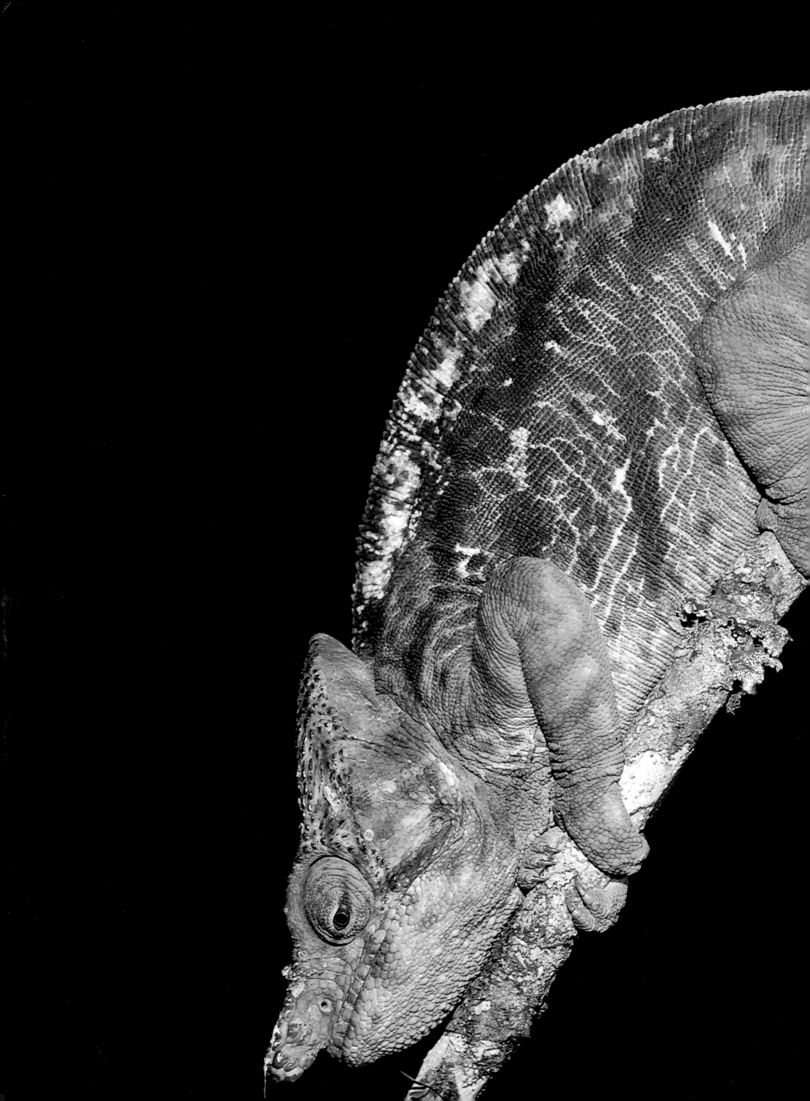

Giant chamaeleon (*Chamaeleo parsoni*), Madagascar.

Foreword

Tropical rain forests are famously tangled, wet and filled with more species of plants and animals than any other habitat on earth. It is less often said that they are also unfailingly full of astonishments, and it is that which over the past 35 years has drawn me back to them over and over again.

Their surprises, however, are not necessarily discovered easily. At first sight, any rain forest seems strangely destitute of animals, bearing in mind all you have been told about their richness. There seem to be none on the ground. You may hear birds calling in the canopy of leaves 45 metres (150 feet) above you, and at times the air is filled with choruses of whistles and chirps, yelps and whirrs that you guess come from either insects or frogs, but there is little to be seen of the singers. Only as your eyes become more attuned are you likely to spot a bird, such as a trogon, sitting motionless on a branch above, gravely watching your every movement; and only if you tread on it are you likely to be aware of the viper that lies curled and camouflaged among the litter of dead leaves on the ground.

As for the trees, they are dismayingly difficult to identify. Cylindrical, branchless trunks rise vertically around you like huge masts, rigged with lianas dangling from their crowns. You soon realize that identifying forest plants is hardly ever, as you may have supposed, a matter of recognizing opulent and glamorous flowers, but more likely a question of studying the finer details of bark. One pioneering botanist working in the forests of Southeast Asia became so frustrated by this that, in desperation, he trained a monkey to clamber up the lianas, break off flowering branches in the canopy and throw them down to him.

But slowly you become accustomed to the place and its apparent uniformity. You develop an eye for detail and start to spot the unusual. Then the astonishments begin. broad drooping leaf of a kind of wild banana in the Costa R forest is occasionally marked by a line of tiny holes on either side of its midrib. Lift it and beneath you find a line of pure white les of fur the size and shape of golf balls. They are nomadic bats that have built themselves a temporary encampment by biting through the side-veins of the leaf so that the two halves flop down and form a neat tent that shields them from the rain. In New Guinea, a shrieking chorus at dawn draws you to a tall tree; and there in the upper branches you see a dozen lesser birds of paradise dancing together, each with a trembling fountain of golden plumes erected over its back in ecstatic display. And the West African forest provides perhaps the most haunting experience of all. A trail of broken stems and crushed leaves through thickets of wild celery and giant stinging nettles two metres (seven feet) high may lead you to a meeting with one of our closest relations, a family of gorillas lounging on the ground as they feed on handfuls of leaves, the youngsters inquisitive, energetic and impish, the mother tolerant and gentle, and the whole group watched over by a magnificent, silver-backed male.

Some of these marvels you may discover for yourself, but you will see much more if you can persuade some of the people who live in these places to guide you. The cost is likely to include serious damage to any pride you might have h n your expertise, either as a seasoned traveller or a sharp-eyed observer. These people move through the forest so much more efficiently

Rain forests are unfailingly full of astonishments, and it is that which over the past 35 years has drawn me back to them over and over again.

than you. They never slip, they never seem to tire, they see things long before you do, and they have to be persuaded that unless they point them out you will not notice animals that they think are obvious. Most remarkable of all, they never seem to get lost. Their knowledge of the forest and its inhabitants is vast and detailed. Their classification of plants is certainly based on quite different criteria from those used by European taxonomy; but their understanding of the uses to which forest plants may be put is the distillation of the experience of countless generations, and is still far beyond the researches of foreign scientists who have only just begun their investigations.

It would be over-romantic, indeed plain wrong, to suggest that all these forest peoples, in addition to their other virtues, are archetypical conservationists, living in perfect harmony with nature. They will often take what they want from the forest with sublime disregard for any principle of conservation. They may fell a whole tree to get a single meal of fruit, or kill a bird for the transient pleasure of putting its plumes in their hair. That the forest is not devastated by such treatment is simply because their numbers are so small and the areas over which they wander are so vast.

Outsiders who intrude into the rain forests can find them uncomfortable, cruel places. Regular heavy rains do not greatly trouble people who wear few clothes and carry little equipment, but a tender-skinned stranger is made wretched by such drenchings and is forced to come to terms with such frustrating occurences as mould growing across camera lenses, and insects eating the insides of books. Lacking traditional knowledge, the stranger has great difficulty in finding food. Leaves that look succulent

(*Left*) Montane forest, Venezuela. (*Above*) Erigpatsca tribesman, Amazonia.

are armed with stings or prove to be poisonous. Fruits hang high up, far out of reach. The animals are so wary that hunting them is extremely difficult. Settlers who move into this strange alien world find after they have laboriously cleared the land that it is not, as they had expected, the fount of fertility. On the contrary, it is so poor that it will produce only one or two seasons of crops before relapsing into near-sterility. Long-distance travellers, whether soldiers or traders, anxious to get from one side of the forest to another, become bogged down in endless quagmires, harassed by clouds of biting insects, and condemned to cross and recross the same river as it meanders its way across their path. So it is not surprising that for centuries the reaction of outsiders, by and large, was either to avoid the rain forests or to destroy them. Some of the giant trees could be sold for profit, but the rest of the vegetation, with its multitude of inhabitants, could be destroyed without any qualms if anyone could think of something useful to be done with the land.

Until recently that attitude had only a marginal effect on the rain forests, for the labour of cutting down the trees was huge and the demands for the land itself relatively small. But over the last half-century there were two crucial changes. The human population began suddenly to increase with great rapidity; and immensely powerful machinery was developed that could bulldoze land clear and flat, and cut down a 100-year-old tree in ten minutes. So the destruction accelerated. Then a decade or so ago, the world suddenly saw that if such a pace were maintained, the forests would disappear totally in the near future. Only at this late stage did we begin to realize what treasures these forests contained and how crucial they were for the ecological health of the planet as a whole.

The alarm at this impending catastrophe was voiced

particularly loudly by people living far from the forests themselves. The smallholders living around the edges of the forests desperately hungry for land, the politicians of rain forest countries wrestling with urgent economic problems that they hoped could be lessened by the quick sale of timber – such people felt considerable outrage at preachings and pleadings from people living far away who had long since totally cleared their own forests and built their own industrial revolution on the proceeds. So Brazilians and Malaysians, West Africans and Papua New Guineans will doubtless be looking closely to see how wealthier nations treat the very few rain forests they have within their own frontiers. The Australians' treatment of their Queensland forests (see page 175) may well be seen as a measure of the sincerity of industrialized nations.

As concern has spread worldwide, so false rumours have started to circulate alongside the only-too-true facts. It is not the case that any form of exploitation damages the rain forest irrevocably, and that any kind of timber extraction must necessarily be disastrous. Although forests worldwide play a key role in storing carbon and releasing oxygen, rain forests alone are not the earth's lungs, and it is very doubtful if replacing them with agriculture and plantations would have a major effect on the all-important oxygen cycle. Nor do we have evidence that hundreds, let alone thousands of species of rain forest animals and plants have been exterminated as a result of humanity's activities – yet. The crisis is too real, too urgently in need of practical solutions to allow exaggerations and illusions such as these to get in the way. The facts themselves are alarming enough.

The rain forests are indeed in great danger, but their loss is not inevitable. Plans to save them must be based on carefully gathered facts. All our knowledge, commercial as well as ecological, sociological as well as geographical, must be deployed to produce programmes that will reconcile the needs of the people who live in the forests and have the most immediate claims of ownership, with those of industrialists and politicians living around their margins, and those of the people in the world at large who, taking a more distant and long-term view, have realized that the tropical rain forests contain some of the world's greatest treasures and are an integral part of both its health and glory.

This atlas, which publishes for the first time accurate maps of the past and present distribution of the forests worldwide, is a crucial step in that direction.

David Attenborough

(*Main pic*) Burned rain forest, Amazonia. (*Above*) Logged rain forest, Sabah.

What are rain forests?

Many people's first experience of a rain forest is from a boat on a river winding slowly into seemingly impenetrable jungle. Joseph Conrad captured the sense of foreboding and mystery of this apparently primeval wilderness in his novel *Heart of Darkness* set on the River Congo (now known as the Zaire): "Going up that river was like going back to beginnings, when vegetation rioted...and the trees were kings."

The term "rain forest" was first coined in 1898 by a German botanist named Schimper, to describe forests that grow in constantly wet conditions. They can occur wherever the annual rainfall is more than 2,000 millimetres (80 inches) and evenly spread throughout the year. Rain forests are found in temperate as well as tropical regions, but the best-known ones, the subject of this book, occur in a belt around the equator. In rain forests the overhead canopy is closed, with few large gaps between trees, a feature that they share with the tropical seasonal forests that grow north and south of the rain forest belt. Tropical seasonal forests are less extensive than the true rain forests, but since they have many features in common they are together often known as "moist forests".

Rain forests can be divided into two broad categories according to altitude – lowland and montane rain forests. Lowland forests are by far the most extensive but, because they are easily accessible, they have suffered the most damage and clearance. They are also the most prolific of all the plant communities in the world. The canopy can reach more than 45 metres (150 feet) in height and consists of many different tree species living close together. A few trees, known as emergents, break through the canopy, often attaining heights of 60 metres (200 feet) with straight unbranched trunks up to 40–50 metres (130–165 feet). The tallest broad-leaf tree ever recorded from the tropics was a specimen of *Koompassia excelsa* from Sarawak, which was more than 83 metres (270 feet) tall.

Montane rain forests are much smaller in stature, their growth restricted by a combination of low temperatures, unpredictable rainfall and the lack of nutrients at higher altitudes. These stunted forests play a key role in protecting the environment; without them, soil erosion in highlands and flash floods in lowlands are the damaging consequences.

Mangrove forest is a type of rain forest found in silt-rich, saline coastal waters. The most extensive mangroves in the world are the Sundarbans in the Ganges delta. Another type of flooded forest is found along the banks of rain forest rivers, where large areas of lowland forest are inundated with fresh water. These include the *igapó* and *várzea* forests of the Amazon Basin, and have been included in the lowland forest on the maps.

Lowland rain forests are the most prolific of all the plant communities in the world.

Attracting birds – In order to attract bird pollinators, many rain forest plants have colourful flowers. The relationship between pollinator and flower is often very specific.

Colourful food – The distinctive fruit of the zebra wood (*Connarus* sp.) found in the Cameroon rain forest advertises its presence to seed dispersers through its bright colour.

Ground cover – The dark, humid conditions found on the floor of the rain forest suit a wide variety of herbs. This herb layer exhibits a large number of different leaf shapes.

Lowland rain forests

Mention the word "jungle" and the image that springs to most people's minds is that of a dark, oppressively hot and humid world with a profusion of hanging vines and creepers concealing any animal life. In reality, it is the lowland rain forest on well-drained soils in which the dense canopy prevents the strong sunlight from reaching the forest floor and also traps moisture. Along river banks and in forest clearings dense undergrowth defies entry into the forest but, like the set of a Hollywood western, it falls away in density behind the façade. Lowland is the most extensive type of rain forest and, as the name suggests, it is found in relatively low-lying areas – generally up to 900 metres (3,000 feet), but up to 1,800 metres (6,000 feet) in western Amazonia (see pages 100–101).

The great Amazon and Zaire River basins are the two largest intact regions of lowland rain forest in the world; most of the lowland forest in Southeast Asia, Central America and West Africa has now been fragmented and altered in one way or another. But parts of the Amazon Basin are now being cleared at an alarming rate, and increasing pressure is being put upon the African forests too.

Botanists have identified as many as 40 different types of lowland rain forest, differing mainly because of unique patterns of rainfall, soil fertility and drainage. Although there are broad similarities in the physical appearance of the rain forests on different continents, the actual species they contain vary widely.

Forests on dry land

In regions with high rainfall, spread more or less equally through the year, the lowland forest is evergreen. Most of the trees do not shed their leaves at all, or do so at different times. In this type of forest, the abundance and diversity of tree species is unequalled in the world, and it is in these rain forests that many fine hardwoods grow. Although neither as tall nor as wide in girth as trees such as the Californian redwoods or Australian giant eucalypts, lowland rain forest hardwoods are highly prized by loggers. Hanging from the trees, and often connecting them at many different levels, are a mass of creepers and vines (lianas), and in the canopy there are many epiphytes (plants that live on other plants).

A few degrees farther away from the equator, the climate becomes slightly more seasonal, rainfall is lowered for one or two months in the year and the forest becomes semi-evergreen in character. Up to one-third of the trees in semi-evergreen forest may be deciduous, but because each species sheds its leaves at its own time, there is no clear season of leaf-fall. In nature, evergreen rain forest grades almost imperceptibly into semi-evergreen rain forest, and the boundary between them has never been mapped precisely. To the casual observer there is very little difference in appearance between the two, but the canopy in a semi-evergreen forest is often not as high that in evergreen forest.

Specialized forests

In some relatively small areas, the presence of a particular type of soil or underlying rock has produced specialized lowland rain forests of various kinds. For example, on free-draining sandy soils heath forests may grow. These are known as *kerangas* in Borneo and as *caatinga* in South America. The plants have to cope with a poor supply of nutrients and often lack enough water for normal growth. *Kerangas* is characterized by an even canopy over a forest made up of many relatively slender trees. In heath forests there is an abundance of plants that have evolved elaborate ways of obtaining extra nutrients. Pitcher plants (*Nepenthes* spp.) are common in *kerangas* and obtain nourishment by trapping insects; ant-plants such as *Hydnophytum* spp. obtain some of their food from ants in exchange for providing a nesting-place in the swollen root of the plant.

Lowland rain forest growing over limestone base-rock has not been studied very much, but it is known to have many unique species. Such "limestone" forests are particularly common in Southeast Asia, and the limestone flora of Peninsular Malaysia

Down by the river (*above*) – A river creates a gap in the forest canopy, a gap that is rapidly seized on by a multitude of young plants struggling for light on the river-bank, creating a wall of vegetation.

Support structure (*left*) – Where the soil is thin or subject to flooding, the trees may gain additional support through the use of buttress roots, which grow from the main trunk at up to 5 m (16.5 ft) above the ground. Where the canopy is relatively open, light can reach the forest floor which encourages the development of a lush undergrowth.

Forest bounty (*above*) – When the fruit of the wild nutmeg (*Myristica fragrans*) is ripe it splits to reveal a crimson-coloured aril which surrounds the seed or nutmeg. Ground nutmeg was used by the Romans as incense, and from about 1600 it became an important commercial spice, being shipped from Indonesia to Europe.

has more than 1,200 species, 130 of which are found nowhere else. Limestone rain forests contain little commercial timber – the growing conditions are generally poor – but they are under constant threat from fire. In Indochina, they have been severely damaged and sometimes destroyed by fire. Even in the much wetter climate of Sarawak, limestone forests on hills are known to suffer fires caused by lightning strikes.

Other lowland rain forest types include those on ultrabasic rocks, beach forests, liana forests and bamboo forests, all of which have their own special features and species.

Flooded forests

Throughout the tropics river levels can rise and fall dramatically, inundating large areas of lowland forest when the rivers burst their banks. Because of the stress caused by waterlogging of the root systems, these swamp forests often contain fewer tree species than well-drained forests. However, few generalizations are possible because they vary widely in terms of the amount of time spent underwater. Several different types of swamp forest are found in all three continental regions.

Freshwater swamp forests were once widespread in Southeast Asia, especially along the banks of some of the region's largest rivers, such as the Mekong and Ayeyarwady in Myanma (Burma); good examples remain the Fly and Sepik in New Guinea. The freshwater swamps of Kalimantan are the natural home of the sago palm (*Eugeissona utilis*) – an important staple food.

In the Amazon Basin, the type of freshwater swamp forest found along the river bank depends on the water. *Várzea* is found on the floodplains of "whitewater" rivers born in the Andes, which carry huge amounts of silt and nutrients into the forest. The forest floor of *várzea* is formed by this sediment, which is trapped by the large buttress roots of the trees, gradually building up the level of the plains. This type of swamp forest is seasonally flooded and found along such whitewater rivers as the Madeira and the Amazon. Below Manaus on the Amazon, the *várzea* is often in a narrow band along the river bank, with grassy meadows inland: upstream of Manaus, the *várzea* is continuous and usually dotted with lakes.

Igapó grows on the sandy floodplains of blackwater rivers, like the Rio Negro, Tapajos and the Arapiuns. Here the water is a clear, dark brown colour caused by rotting plants, but it contains little suspended material. The sandy soil and clear water do not allow the soil to build up in the same way as it does in the *várzea*. In the dry season, it is usual to find sandy beaches with trees growing out of them. *Igapó* is usually flooded for between four and seven months each year, up to a height of 12 metres (40 feet). This does not kill the trees; life continues above the water level.

Palms are often the dominant species in freshwater swamp forest of the Amazon, but trees such as the kapok or silk-cotton tree (*Ceiba pentandra*) with its giant buttressed roots, often form the main canopy. Most of the flooded forest plants flower in the dry season when the water is low. However, the kapok flowers in the high-water season, releasing cotton-like seeds that are blown in the wind and eventually swept away by the river.

Low-growing plants, particularly in the *igapó*, can be submerged for most of the year, enjoying only a brief spell above water. They flower and grow fruit in this time, and capture as much sunlight as possible before being inundated once again. Saplings and seedlings can spend their first 20 years only briefly seeing daylight above the water.

Unlike freshwater swamp forest, which is regularly or occasionally flooded, peat swamp forest, when fully developed, rises above the flood water level. Peat swamps are present in some parts of the Amazon Basin, and on some Caribbean islands. They are rare in Africa, but common in Southeast Asia, covering 12 percent of the islands of Borneo and Sumatra, as well as large areas in Irian Jaya.

Flooded lowland rain forest, Amazonia.

In peat swamps the solid fibrous crust, which is somewhat spongy to walk on, covers a semi-liquid interior and the drainage waters are a rich dark colour. Plant growth on the peat is often zoned concentrically, the centre zone supporting the growth of very stunted plants with thin trunks, whereas the outer zones are dominated by larger trees which can reach up to 50 metres (160 feet) in height. In Sarawak on the island of Borneo, some of the peat swamps have up to six recognizable zones.

Variations around the world

Throughout the world, rain forests share some general features, such as buttress roots and an abundance of lianas and epiphytes, but floristically they are almost totally different from one continent to the next. The tree *Symphonia globulifera* is one of very few rain forest species common to both the African region and South America, and the Bromeliaceae, an important family of epiphytic and other plants in South America, has only one species in Africa. Despite the fact that these two continents share so few species, they have more in common at the family level. Both have many species of the major tree families – Meliaceae (mahoganies), Sapotaceae, Euphorbiaceae and Leguminosae. In contrast, the rain forests of Southeast Asia are dominated by giant trees from the family Dipterocarpaceae, which are almost absent from Africa and South America. Dipterocarps flower very irregularly, probably triggered by droughts at intervals of between five and nine years. In a "dipterocarp year" the forest canopy becomes a mass of colour as one species after another comes into flower.

Species diversity seems to be highest in the South American rain forests, where at least 30,000 species of higher plants occur, and lowest in Africa (including Madagascar) which has about 17,000. The forests of Southeast Asia include about 25,000 species of flowering plant, but there are fewer farther east towards the Pacific, where only about 4,000 species are found.

Local species densities are also often impressive, and ecologists working in the rain forest regularly break each other's records. The highest recorded density of species is 233 in only 100 square metres (120 square yards) found in a Costa Rican rain forest. Other impressive records include 1,025 higher plant species in 170 hectares (420 acres) of rain forest in coastal Ecuador, and 350 species in half a hectare (1.2 acres) in Ghana.

Moving away from the equator

In tropical regions where rainfall is more markedly seasonal and there are three or more dry months each year, the tropical forests become deciduous in nature, shedding their leaves during the dry season. In Asia, they are known as "monsoon forests", since they come into leaf during the onset of the annual monsoon, and they are most widespread in parts of Indonesia. In both Africa and Southeast Asia there is very little seasonal forest left; most of it has been both cleared and burned in the past by humans. In many parts of Africa, seasonal forest has been degraded to savanna woodland which has a grassy floor and grows right up to the edge of the rain forest. In South America, forests growing in seasonal climates tend to hold their leaves longer and some are even evergreen, which makes them less prone to fire damage.

One of the main factors that differentiates seasonal forest types from true rain forest is the lack of climbing plants which link the trees together. These cannot survive in the drier air, and their absence serves further to reduce the amount of moisture in the interior of the forest. During the dry season, the trees lose their leaves, increasing the amount of light able to reach the forest floor, enabling many more plants to grow there. Many trees flower in the dry season, their colours exaggerating the leaflessness of their surroundings.

Human impact on the forests

Human activity in the lowland forest does not necessarily mean that it is permanently cleared away and converted to other uses. After being logged or used for agriculture, the forest may be allowed to regrow into what is often called "disturbed" or "secondary" forest. These forests are arguably even more variable than the more natural forests in which they have their origin. Many authors use the term "disturbed" (sometimes "degraded" or "logged-over") to describe those forests that have been subjected to logging, while the term "secondary" is reserved for those that have regrown after complete forest clearance (usually by shifting cultivators). Disturbed or logged-over forests can be extremely difficult to identify, even for an expert, after a few decades of regrowth. Of course, this depends upon the original intensity of logging. In Central Africa, where only one tree may be removed per hectare, very little damage is done; in Sabah or the Philippines, however, most of the trees are either removed or destroyed in the logging operations.

Secondary lowland forest is full of fast-growing "pioneer" species that have invaded the abandoned farmland, and is floristically quite different from its more mature antecedents. This regrowth is in fact more "jungle-like" than primary forest, due to the mass of herbs, shrubs, lianas and young trees that cover the forest floor. In contrast, it is relatively easy to walk through a stretch of primary lowland forest since the interior of the forest is so dark that little grows on the forest floor.

Roots (*above*) – The floor of this rain forest in Costa Rica is criss-crossed with buttress roots which stabilize the trees and soil.

Popular stimulant (*left*) – The seeds of the guarana (*Paullinia cupana*) contain about 3 times as much caffeine as coffee. They are used to produce a tonic that is sold throughout South America.

Dipterocarp forest (*right*) – From a high vantage point it is possible to see tall dipterocarps rising above the canopy in this Malaysian forest, creating a characteristic billowing appearance.

Mangrove forests

The shores of the tropics often support mangrove forest, a special type of rain forest which is poor in its variety of species. Mangrove forests are the normal plant communities of sheltered tropical shores, but they can extend to about 32°N and even farther from the equator in the southern hemisphere. The biggest and richest mangroves are in the wet tropics where the rain forests are found, particularly on the coast of Bangladesh, the Malay Peninsula, Sumatra, Borneo, New Guinea and the other islands scattered throughout Southeast Asia.

Mangroves are evergreen trees and shrubs which although they belong to several unrelated families share a similar habitat, namely silt-rich soils in saline coastal waters. These plants show a number of adaptations to the salty environment including pneumatophores or breathing roots. These emerge from the waterlogged mud into the air, where they can absorb oxygen needed by the root system. Such roots can also occur in other types of waterlogged forest, particularly freshwater swamp forest. The breathing function allows the tree to live in soil almost totally deficient in oxygen, but they also have another purpose. In mangrove forests the soil level is constantly rising and the pneumatophores enable the tree to produce a fresh crop of rootlets at successively higher levels. The absorptive part of the root system is kept near the surface of the soil while the deeper roots anchor the tree into the ground.

Mangrove forest can be as much as 30 metres (100 feet) tall or be limited to a poor shrub only a few metres high. On the edges of a mangrove forest, where the trees either stand in the sea or on mud flats which only dry out at low tide, the roots provide a backdrop for large numbers of birds, crabs and molluscs. Moving deeper into the forest, the trees become higher and more impenetrable.

Mangroves around the world

The world's mangroves can be broadly divided into two groups; an eastern group on the coasts of the Indian and western Pacific oceans, and a western group on the coasts of the Americas, the Caribbean and West Africa. They share many similarities, but the eastern forests have a greater variety of species. All the genera of the western group are found in the east, but the species are different. Only Fiji and the Tonga Islands, in the Pacific, have an eastern as well as a western species, *Rhizophora mucronata* and *Rhizophora mangle*.

The biggest mangrove forests are the Sundarbans of the Ganges Delta, which straddle the borders of India and Bangladesh. In Asia, mangrove forest has been extensively cleared for development of fish and prawn pools. Most of the mangroves that were once commonly found around many of the Philippine islands have now been removed for this purpose. Less permanent degradation results from the use of mangrove bark for tanning, trunks for building and branches for firewood. Of particular concern is clear felling of Asian mangrove forests for the production of paper in Japan. In particular, Irian Jaya's large mangrove forests are beginning to attract the attention of paper companies worldwide.

The Amazonian mangrove forests occur only in a narrow coastal belt, and only the larger areas have been shown in the map section. Speciation is again poor, the major species being *Rhizophora mangle*, *Avicennia nitida*, *Laguncularia racemosa* and *Conocarpus erectus*. Whereas true mangrove – *Rhizophora* – only extends as far as the influence of salt water, white mangrove or *Avicennia* winds deep inland into freshwater areas. *Rhizophora typica* is widespread in the mangroves around Aracaju and Recife on Brazil's Atlantic coast, extending up rivers as far as 20 kilometres (12 miles) inland. *Rhizophora racemosa* is more common near Marajó Island in the mouth of the Amazon and from there to Guyana. Mangrove forests are also found on the north coast of Brazil, between Maracá Island and the border with French Guiana, a muddy region with few sandy beaches caused by the marine current that carries the sediment further west.

Designed to stick (*above*) – The seeds of some mangroves are long and pointed so that when they drop from the parent plant they stick into the mud below, thus reducing the chances of being washed away.

Mangrove roots (*left*) – Among the tangle of roots that anchor the mangrove, small finger-like breathing roots (pneumatophores) stick up through the mud and young saplings grow near to their parent.

The mangrove edge (*right*) – On the coast of Queensland in Australia, 2 types of rain forest can be seen side by side. Twisted roots characterize the mangrove forest at the water's edge; inland, lowland rain forest rises up from the shore.

Montane rain forests

At higher altitudes on forested mountains in the tropics, the hot, sticky humidity of the lowland rain forest gives way to a cooler dampness. As the climate changes, so do the flora and fauna: above 900 metres (3,000 feet), the species found are usually rather different from those lower down. These high-altitude or montane forests are generally further subdivided into recognizable bands: lower montane forest (often in the 900–2,000 metre (3,000–6,600 feet) range), upper montane forest (in the 2,000–3,200 metre (6,600–10,500 feet) range) and subalpine forest (which may extend as high as 3,800 metres (12,500 feet)). Mists often engulf the forest canopy, accounting for the evocative name "cloud forest". Light levels are greatly reduced and the foliage literally drips with water that condenses out from the air.

The canopy drops rapidly at higher altitudes. In the lower montane formations 15–33 metres (50–110 feet) is usual, whereas in the upper montane forest it may be reduced to little more than the height of a person. Features of the lowland forests, such as buttressed roots and fruits growing on tree trunks, disappear. Lichens hang like beards from the uppermost branches, and the trunks and forest floor are covered in bright green mosses, oozing water. Filmy ferns and other epiphytes, including orchids and bromeliads, become quite common. Trees lose the straight, unbranched trunks of the lowlands and become gnarled, twisted and often multi-stemmed. The tops of the trees are no longer irregular and billowing, but more even and flat. Even the leaves are much smaller, narrower and more leathery.

In the tropics the average temperature falls by between 0.4–0.7°C (0.7–1.3°F) with every 100 metres (330 feet) gained in altitude. Heavy rainfall and less evaporation, due to the low temperatures, combine with strong winds to restrict plant growth. Decomposition is slowed down, as well as growth. Deep layers of waterlogged peat accumulate in valleys. Termites cannot survive up here, but the soil contains large earthworms and beetle larvae. As nutrients are only slowly released, specially adapted plants, such as the pitcher plants (*Nepenthes* spp.) occur here.

Most lowland species cannot survive at these altitudes, and specialist trees more reminiscent of families in the temperate latitudes, such as oaks, beeches and laurels, become more common. Overall there are far fewer species than are found in the lowlands, but many cold-adapted groups come into their own. Giant heathers and rhododendrons from the family Ericaceae occur on mountain-tops in Asia. Conifers, generally found only in nutrient-poor conditions in the lowlands, become more common at altitude, particularly in Asia, where species of *Dacrydium* and *Podocarpus* are prominent. *Podocarpus* forests are also found in Colombia, Peru and Venezuela, but large areas have been cleared. *Araucaria* species become widespread at high altitudes in the Andes of southern Brazil, but these are also important timber trees and are now endangered in the wild.

In Africa, montane forests are scattered over the continent, notably in Ethiopia, the highlands of the Rwanda-Burundi-Zaire border, Kenya, northern Tanzania and Cameroon. Bamboo is widespread at between 2,500 and 3,000 metres (8,200–9,800 feet), and coniferous junipers (*Juniperus* spp.) and *Podocarpus* forests occur higher up. Many isolated montane forests have their own unique assemblages of plants and animals, and are of particular conservation importance.

A survey of Malaysian flowering plants revealed that nearly every plant family was restricted to a certain altitude range. However, similar studies in New Guinea have shown that many species span the whole range of altitudes, although with increasing altitude the plants are smaller, and some become multi-stemmed, with smaller, mature leaves.

The distribution of animals is also affected by altitude, depending on their eating habits and differing lifestyles. In general, the number of omnivores usually does not decline with altitude, although fruit eaters, predators and scavengers become less common higher up. The number of insect, bird and reptile species also drops with altitude.

Up in the clouds (*main pic, insets*) – Mosses and lichens flourish in the cool humidity found in montane rain forests, covering both the trunks of trees (*above right*) and the tangle of climbing plants and lianas (*main pic*). There are also many fungi (*above left*), including mushrooms, toadstools and moulds.

Why we need rain forests

Throughout history, people have cut down trees, and converted forests into land for farming and other such "productive" uses. For example, much of Europe was densely forested 1,000 years ago. Converting these forests to other land uses has fuelled the phenomenal social, economic and technological development that has ensued. Now tropical countries are developing rapidly. If they follow the model of the industrialized northern nations, why should they not cut down their forests, clear the land, and turn it over to agriculture or industry? Together, tropical rain forests cover a land area the size of the United States. Is this not just wasted land waiting to be used?

Such arguments are not only out of fashion these days, they are also wrong. The rapidly "greening" consumer society in the affluent northern nations is rallying behind environmentalists to preserve the rain forests, and there is a sound scientific basis to this movement. But what is it about rain forests that causes such strong feelings? Do we really need them at all?

It is vital to realize that tropical forests are *qualitatively* different from temperate forests. We have been able to cut down huge areas of temperate forest and convert the land for agricultural purposes without any obvious detrimental effects. In a sense, clearing the temperate forests has been a precursor to development; clearing tropical rain forests, on the other hand, could be a precursor to disaster – both for the people in the developing tropical nations, and the entire human race.

The point is that people – all people on earth – need tropical rain forests. At the most fundamental level, rain forests provide a home for millions of tribal people, who have adapted to life in this unique habitat (see pages 90–95). Surely they have a right to continue living in their traditional lands. For them, the forests provide shelter, animal and plant products, and food. In short, their whole livelihood rests there.

But many other people, who do not actually live in the forests, rely on them just as much as the tribespeople. Rain forests appear robust and impenetrable, but in reality they are ecologically very fragile. Most rain forest soils are almost infertile, being poor in nutrients and susceptible to erosion. Lose the tree cover, and the root systems that hold these forests together, and the little fertility that remains, or even the very land itself, may be lost. Rains in tropical regions are not like the gentle drizzles of temperate climes; instead they come in short, sharp downpours which leach out the nutrients in the soil and quickly erode exposed topsoil. Tropical forests break the impact of these sudden storms. Not only is the soil protected, but the water supply is regulated. Forests receive intermittent doses of heavy rain, but give out a steady supply of water, which is currently taken for granted by millions of people downstream. They also play a crucial role in regulating local rainfall patterns.

The fact is that most tropical soils can sustain only slow-growing trees; only relatively small areas are fertile enough and stable enough to be suitable for agriculture, a situation fundamentally different from that in more temperate latitudes. In addition, the real effects on the world's climate of burning the last great reserves of organic carbon can only be guessed at.

As if all this is not enough, cutting down rain forests may have other equally important consequences. Rain forests differ from temperate forests in the sheer range and diversity of life that thrives under the canopy. It is now known that although they cover less than about six percent of the earth's land area, they contain more than 50 percent of all species. It is essential to preserve this biological diversity. The genetic resources contained in the forests are the common heritage of humankind, and may well prove to be vitally important to the future welfare of the human race. Already many important medicines and drugs are derived from plant species unique to rain forests, and scientists believe that many more will be discovered. We are in danger of finding that, just when the lid on Nature's medicine chest is being opened, we have lost the contents. Extinction is for ever.

The point is that people – all people on earth – need tropical rain forests.

For[...] [...] **andless settlers** – For those
live w[...] [...]aping from the poverty in
the Kayapó Ind[...] [...]n slums, forested land can
Amazon Basin, then [...]sed to start a new life. But
provide all that they need [...]th little knowledge of their
including the materials to make [...]ew environment, they are
a hammock. [...] often unsuccessful.

City dwellers (*main pic*) – [...]e populations of large urban
centres, both in the indus[...]lized world and the Third
World, rely on the rain forest[...]ot only for such products as
timber and medicines, but a[...]so for a stable climate.

The human factor

Rain forests have long provided a home for people. Scientists have unearthed 12,000 years of human habitation in the Amazon Basin, and 39,000 year-old artefacts have been excavated from caves in the Bornean jungle. Today, it is estimated that as many as 50 million tribal people may still live within the world's tropical forests. For these people the forest is their home, their spiritual base, and the source of their food and clothing.

Generally these hunter-gatherers (see pages 92–93), hunter-gardeners or shifting cultivators (see pages 94–95) live in small communities, often with highly developed social and cultural systems. But most importantly, these people know and understand the importance of the forest to them as the source of their whole livelihood. Some have developed remarkable ways of using it in a sustainable way for their needs.

Although tribal people most obviously need rain forests, they are not the only ones for whom these forests provide a livelihood. The *seringurios* or rubber tappers of Brazil (see page 124) rely on rain forests for their living. The murder in 1988 of Chico Mendes, a former president of the rubber tappers' union, has done much to publicize their plight – and underlined the point that the forest is needed intact, as a valuable economic resource.

Very few rural settlers realize the importance of maintaining the forest intact. Nor do they understand how to manage it in a sustainable way. Throughout the developing world, unequal land ownership is forcing people either to head for the cities, or to carve out a new life in the forests by using the roads made by logging and mining companies.

Wood as timber and fuel

Tropical forests are also important to more than two-thirds of the people living in developing countries (mainly the rural poor), who depend on wood for their household energy needs. In several parts of the world where population pressure is heavy, there is an impending shortage of fuelwood as forests are hacked away. Sub-Saharan Africa, parts of China and the Indian subcontinent are particularly badly affected. Collection of fuelwood (normally the task of women) is beginning to dominate the lives of millions of rural people, taking up an increasing proportion of their time.

This is an issue that relates more to the drier seasonal forests and woodlands, rather than wet rain forests – mainly because population pressure is not yet as great in or around the existing rain forests. But it highlights the importance of all forests as a source of wood, and makes the wastage of wood, either through burning or just leaving whole trees to rot away, all the more frustrating.

To people in the timber trade, rain forests are an important source of fine hardwood timber. The international tropical timber trade, now a multi-million dollar business, is rapidly realizing that if the forests die, so does the trade; and none of the produc countries, the merchants, or consumers wants that. Only commercially logged forest can be maintained as forest is the any hope of achieving a long-term supply of the hardwoods prize by furniture-makers, boat-builders and interior designers, because it is not possible to produce the same range of tropical hardwoods in plantations.

Ultimately, rain forests are important because people need them. Those who have adapted to life within the forests require them for their everyday needs; and responsible logging compa nies need the timber, but realize that forest cover must remain intact if there is to be any hope of a sustainable timber harvest.

The threat to the forests continues to be the competing needs of burgeoning populations of agricultural colonists, and the greed of timber merchants looking for a quick profit. The plight of the colonists attracts some sympathy – their lot can be improved only through changes in government policy and land-tenure laws – but careless commercial logging is inexcusable, given the profits that timber companies make. It remains to be seen whether the competing needs of tribal peoples, the rural poor and the commercial logging companies can be resolved.

Fuel for fires (*above*) – The Yanomami sometimes travel long distances to collect wood to use as fuel for cooking. Although some of their diet comprises raw vegetable matter, a large proportion is fresh meat (including fish, monkeys and wild pigs) which has to be cooked.

Rubber (*right*) – To collect latex, the raw material for rubber, a number of slanting cuts are made in the trunk of the rubber tree (*Hevea brasiliensis*). The latex oozes from these cuts and runs downwards into a small cup attached below.

Made in the rain forest

A recent study by a team headed by Dr Charles Peters, of the Institute of Economic Botany in New York, claims that fruits and latex represent more than 90 percent of the total market value of the section of Amazonian forest they studied: "The results from our study clearly demonstrate the importance of non-wood forest products. These resources not only yield higher net revenues per hectare than timber, but they can also be harvested with considerably less damage to the forest."

Forest products such as nuts, fruits, rubber and rattan all grow naturally and are harvested locally, but they have never been seriously considered by economists as part of the overall commercial value of a rain forest. Peters suggests that this is because timber is a high-profile export sold in international markets, and thus is highly visible. Non-wood resources are collected and sold in local markets by a large number of local people; their value is hard to monitor and easy to overlook. From the point of view of the government of a developing country, timber is more easily sold for hard currency which can be used to service international debts.

The idea that rain forests can produce more than timber is not a new one in the tropical world. Indeed, it has been the basis for local economies in the forests for thousands of years. The bias towards timber as the only worthwhile commercial product may stem from the fact that the methods used by logging companies originated in Europe. There are qualitative differences between tropical and temperate forests that are still not fully appreciated in Europe, North America and Japan: in the forests in these regions there is very little commercial value beyond the timber, whereas tropical forests are far richer in other potentially valuable products. Overlooking the non-wood products may be a case of the application of Western thinking and techniques to tropical forests in inappropriate ways.

New values

The new thinking is towards using tropical rain forests as "extractive reserves", from which a wide range of products can be harvested on a sustainable basis, including some timber. But the only way in which this will become a viable reality is if a market for such products is created in the industrialized world. Professor Ghillean Prance, the Director of Kew Gardens in London, has called on Western companies to create and develop just such markets, and several are already investigating new ways of using non-wood forest products.

Some of the nuts and other rain forest products are already familiar, but others are entirely new. For example, Peruvians are used to eating ice cream coloured and flavoured with the purple fruit of the mauitia palm. A company in the United States is now marketing "Rain forest Crunch" – an ice cream made with brazil nuts gathered wild and cashews from replanted areas.

In the United Kingdom, one "green" company is looking into the marketing of a wide range of forest products, ranging from aromatic bath oils and body creams to new pot-pourris. Resins, latex and additives for cosmetics are being investigated. But the scope is even more wide-ranging than this. At least 1,650 known tropical forest plants have potential as vegetable crops, and the sap of the Amazonian tree *Copaifera langsdorfia* is so similar to diesel fuel that it can be used in truck engines.

Rattan is a non-wood product that is already used extensively for furniture manufacture; it grows as a creeper on trees in the forest. Currently rattan fetches much more per tonne than does timber. Rattan can be sustainably harvested from the forest – provided that the forest is maintained intact. The world rattan trade is worth about US$2 billion annually – but is now under threat in Southeast Asia because so much forest has been lost.

If non-wood forest products can be viewed as viable high-value exports, rather than just entities traded at local markets, governments of developing tropical nations will be much more inclined to safeguard the future supply of these products. Safeguarding supply of these implies safeguarding the forests.

Palm weaving (*top*) – The Kanela Indians of northeast Brazil use local palms for weaving baskets and mats which they use in their everyday lives; these and other artifacts may also be used to barter with other nearby communities.

Fruit harvest (*above*) – In the state of Rondônia in Brazil, the Urea-Wau-Wau Indians collect many different fruits from the rain forests including those of the pupunha palm (*Bactris* spp.). After cooking, the flesh of the orange fruit is edible and highly nutritious. These and many other non-wood forest products are vital to the Indians.

Apiarist at work – The knowledge of the local people includes their method of collecting honey, using a specific type of moss that burns very slowly, gives off clouds of smoke and stupefies the bees. First the moss has to be harvested and very carefully set alight.

Passing on knowledge (*left*) – The chief of the Oku tribe of the Kilum highlands holds the plants and plant products of the local rain forest. It is his responsibility to ensure that the tribe's intimate knowledge of the forest vegetation is passed on to the Kilum Project.

Local symbol (*above*) – Bannerman's turaco (*Tauraco bannermani*), which is featured on the Kilum Project logo, is found only in the montane forests of this part of Cameroon. Its head feathers are worn by local tribal chiefs as a symbol of their authority, and its local name, *fen*, has been adopted as the name of the Project's newsletter.

Kilum: making the forests pay

Few examples exist of forests that are being exploited sustainably. Yet in the remote Bamenda highlands of Cameroon a development project is demonstrating that this is possible. The Kilum Project – launched by the International Council for Bird Preservation in 1987 – is, first and foremost, a conservation project. Its 120 square kilometres (45 square miles) of surviving forest contains the best *Podocarpus*/bamboo forest outside East Africa, and shelters a number of rare and threatened species. But John and Heather Parrott, project coordinators, realize that conservation of the forest is impossible without ensuring some benefit to the local people who live beside the forest and who have traditionally looked upon it as a provider of food, medicine, wealth and cultural resources.

By tradition the Kilum region is famed for its numerous species of medicinal plants and the skill of local people in their use. The Project is thus encouraging managed exploitation within the reserve, and traditional doctors are issued permits for the gathering of medicinal plants. Similarly, the felling of trees for the carving of traditional artifacts for cultural or religious purposes is controlled by strict licensing laws, and stipulates regular replanting within and around the forest. Various related forest-based industries are also being evaluated, surrounded by equally strict regulations, such as the production of high-quality hand-crafted paper from the inner bark of *Lasiosiphon glaucus*, common on the forest's edge. The sole occupation of many households, on the other hand, is the collection of honey from hives in the forest: this too is a sustainable resource. Recognizing in both cases the economic potential in these occupations, the coordinators have encouraged the creation of honey and handicraft cooperatives. Production of honey is now approaching 500 tonnes (440 US tons) a year.

According to John Parrot, "with the extraction of any forest product it is essential to monitor the level of exploitation to ensure a sustainable yield." If properly managed, he believes, the potential is immense. Yet the basis for any long-term industry lies with the local people who possess such intricate knowledge of the forest.

The world's genetic library

The diversity of life in a rain forest is truly astounding. At Yanomono, near Iquitos in the Peruvian rain forest, Alwyn Gentry, a botanist from Missouri Botanical Garden, noted 300 species of tree with a trunk diameter of ten centimetres (four inches) or more. Compare that with a temperate forest, which usually contains about a dozen different species per hectare. Rain forests are also buzzing with animal life. Recent research suggests that they could hold as many as 30 million different species of insect.

In terms of sheer biological diversity, Latin America is probably the richest region, followed by Southeast Asia. For example, Colombia possesses an estimated 25,000 indigenous plant species – as many as are found in the whole of Southeast Asia, an area four times larger than Colombia. The United States, by comparison, contains about 17,000 plant species.

There are great variations in diversity not only between different continents, but also at a local level. Forest type, altitude, rainfall, humidity and temperature all make a difference to the type and number of species found. The most diverse areas are those which survived the last Ice Age. During this period, much of the Amazon Basin was transformed into savanna. Only sixteen patches of forests survived from earlier times. Today, these areas, along with similar ones in Africa, contain exceptional biological diversity (see page 50) and should be conserved as a global heritage.

The rain forest is a seething interaction of myriad species, which together make up the ecosystem. However, such a complex system is very delicately balanced. It may take only a relatively small change – for example, a slight climatic alteration, or interference by humans – to disrupt the whole system and send hundreds of species to extinction. Some scientists fear that up to 50 species become extinct each day due to rain forest clearance.

Making use of the genes

"I believe that the world is a poorer place for each species that we lose," said author and naturalist Gerald Durrell. And he is right in more than one sense: extinction is not just a moral issue, involving responsibility for the loss of individual species; each extinction also represents a loss of unique genetic material. The tropical rain forests are a storehouse for more than 50 percent of the world's genetic material, and the importance of maintaining this material cannot be overemphasized.

In agriculture the range of different crops planted around the world has been reduced dramatically in recent times, even as yields increased on average by at least 100 percent between 1930 and 1975. About half of this increase can be credited to genetic improvement and cross-breeding. Globally, we now rely on just eight crops to provide 75 percent of the world's food. This lack of diversity renders us extremely vulnerable to foodcrop pests and diseases, and climatic change. In future, wild plant species may prove vital to adapt current varieties to new conditions.

Botanic gardens and gene banks around the world provide safe havens for a wide range of plants. In 1985, the World Wide Fund for Nature and IUCN set up the Botanic Gardens Conservation Secretariat, which aims to expand the role of botanic gardens in conserving threatened plant species. Today the Secretariat has more than 200 members, including gardens in Latin America, Africa and Southeast Asia. Scientists have a mammoth task ahead, to identify and rescue as many new discoveries as possible before possible extinction.

At the moment, fewer than one percent of tropical rain forest plants have been chemically screened for useful medicinal properties. Yet more than 1,400 species are thought to have anti-cancer properties, and many rain forest products are vital to today's pharmaceutical industry. An average of one in four of all purchases from high-street chemists (drug stores) contains compounds derived from rain forest species. But scientists believe that we are only scratching the surface.

From forest to laboratory – It is the vegetation on and just above the nutrient-poor soil, that is most rich in potentially useful plants (*main pic*). To store the seeds of these plants for future research, refrigerated seed banks (*inset*) are used.

Protecting our environment

The environmental effects of destroying huge tracts of rain forest are something frequently discussed, but not always well understood. All forests are an integral part of the earth's life-support systems, and play an important part in regulating climate and hydrological cycles, as well as maintaining and conserving soils. But, because of the unique nature of rain forests, the environmental effects of clearing them are much more damaging than those experienced after forest clearance in temperate lands.

Tropical rain forest nations are cutting their own throats by cutting down their forests. Deforesting hillsides allows the heavy tropical rains to wash away whatever thin and fragile soil there is. A study in Côte d'Ivoire showed that the annual loss of soil on a forested slope was 30 kilogrammes per hectare (160 pounds per acre); by contrast, a similar deforested slope lost a staggering 138 tonnes (121 US tons) of soil a year.

Deforestation is also responsible for flooding and droughts in many tropical forest countries. Rain forests, with their thick foliage and complex root systems, regulate water supplies. Typically, in a well-forested watershed, 95 percent of the annual rainfall is detained in the sponge-like network of roots in the soil. A lot of this water is released back into the atmosphere by evaporation and transpiration (the process by which water is drawn up from the roots of a plant and evaporated from its leaves), thus reducing the total water run-off, but the remainder is released slowly throughout the year, keeping streams and rivers flowing even during dry seasons.

Globally, more than one billion people depend on water from tropical forests for drinking and crop irrigation. Without the regulatory function of the rain forests, heavy rains result in floods and landslides, whereas rivers dry up if the rains are poor. Floods in Thailand in December 1988 cost 450 lives and caused damage to property worth hundreds of millions of dollars. The Thai government immediately imposed a total logging ban after it became apparent that the floods were almost certainly caused by logging and the clearing of steep slopes for rubber plantations.

Rain forests also affect the local climate. They are fundamental to maintaining rainfall patterns by returning huge amounts of water to the atmosphere. An Anglo-Brazilian study in this field has found that 10–20 percent less water evaporates from cleared areas than from forested areas. Cutting down trees therefore reduces atmospheric humidity and so reduces rainfall.

The climatic effects of tropical rain forests are so powerful that they are felt thousands of miles away from the tropics. By pumping enormous quantities of water into the atmosphere, they have a cooling effect in the tropical regions, and act to warm the higher latitudes. There are two effects at work here. First, the clouds generated over the forests reflect sunlight away from the tropics. Second, evaporation cools the leaves of the trees and as the water vapour condenses in the clouds above the forest the heat is regenerated. Because the circulation of the air masses is away from the equator to the higher latitudes, a proportion of this heat is transported outside the tropics to cooler latitudes.

The animal and plants of the rain forest, just like all other living organisms, are based on carbon (see page 57). As rain forests are such a concentration of life, they contain huge amounts of carbon. When they are burned, this carbon is released into the air as carbon dioxide (CO_2), which is one of several greenhouse gases that occur naturally in the atmosphere and help to regulate temperatures on the earth's surface. During the last hundred years, the amount of CO_2 in the atmosphere has increased steadily. Most of this increase can be accounted for by the burning of fossil fuels (coal, gas and oil). These fuels are essentially carbon compounds created thousands of years ago from dead plants and animals. Currently, about five billion tonnes (4.4 billion US tons) of carbon as CO_2 are pumped into the atmosphere from chimneys and exhaust pipes. The contribution made by the burning of forests is much harder to calculate, but may account for a further one billion tonnes (0.88 billion US tons) a year.

The only short-term way of decreasing the concentration of atmospheric CO_2 is to plant more trees. As a tree grows, it absorbs CO_2 and incorporates the carbon into its cells. When mature, it is in balance with the atmosphere, releasing about the same amount of carbon dioxide into the atmosphere through respiration as it absorbs through photosynthesis.

Large-scale deforestation in the tropics thus threatens to change global climatic systems, by altering the mechanisms by which heat is transferred to higher latitudes. No one can predict with any certainty the outcome for the global climate if this were to occur. If deforestation continues at the present rate the amount of forest burning will increase, adding to the greenhouse effect.

Recycled rain

A research team headed by Aneas Salati at the Brazilian Space Research Institute has concluded that nearly 50 percent of the rain falling over the Amazon Basin is returned to the atmosphere from the forest. In this way, the recycling of water from east to west across the Amazon Basin plays an important part in keeping the Amazon Basin wet. The westerly regions are thousands of kilometres from the Atlantic Ocean, and rely on water "passed on" through the forest-atmosphere system. If large areas of forest in the east are destroyed, this recycling "conveyor" could break down, and lead to a gradual drying out and ultimate death of the forests in the far west of the Basin.

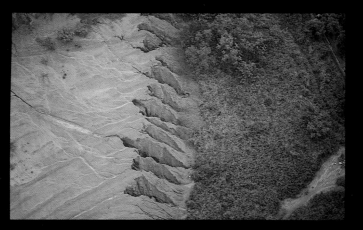

The naked edge (*above*)   From the air, the effects of deforestation in the Amazon Basin are only too evident. Exposed to the elements, the irreplaceable soil is washed away by the heavy tropical rains. Soon, deep gulleys form which increase further the erosive effect of the rainfall.

Pressures on the rain forests

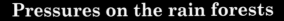

Rain forests offer a wide array of resources in areas of the world that are frequently beset by poverty and rapidly expanding populations. Demographers predict that by the end of the next century four out of every five people in the world will be living in a tropical forest country. Despite an already wide spectrum of wealth in these countries, all are striving for a better standard of living. Some of them – Malaysia, Brazil and Indonesia – are now at various stages of industrialization, whereas others, notably in tropical Africa, are still struggling to find the path to development. During the next century there will be increasing pressure on forests caused by the needs of the people in these countries. Like us, they will need homes, roads, food, educational and recreational facilities, electricity and industrial consumer goods. In many ways, the natural resources of the forests are the raw materials for this development.

One of the most pressing needs will be for agricultural land, but the resources in the forests go much deeper than this: rivers can be dammed for electricity generation, minerals under the forest can be exploited for industry, and the forested areas themselves can yield many different forest products and timber (see pages 30–31), as well as being a genetic storehouse (see pages 32–33).

Much of the land clearance during recent decades has been haphazard, taking place spontaneously without any form of control or planning. Landless settlers have hacked down and burned countless thousands of square kilometres of either undisturbed or logged forest. They gain access to the land by using roads built by governments, or by logging or mining companies, into previously inaccessible forest. The result has frequently been severe environmental degradation. All too often, the type of shifting cultivation that they use is unsustainable.

The ecosystem is destroyed for ever, and the cleared land has to be abandoned after a few years. The settlers can only move on to repeat the process elsewhere.

With the ever-increasing population pressure in tropical forest countries, it is inevitable that this type of deforestation will continue. But these people are no more the root cause of the destruction than soldiers are of wars. Poverty, population growth and unequal land ownership are the fundamental causes of this *ad hoc* land-conversion and destruction. It is this sort of wastage that cannot be allowed to continue. The tropical forest resource is still very large – particularly in South America and parts of Central Africa. But it is not infinite; once trees have been cut down, and the soil has been eroded over very large areas, there is little chance of the forest ever regenerating satisfactorily.

But it is not only the poor who are placing demands on the rain forests. It is the rich industrialized countries that provide the demand that drives the tropical timber trade and the markets for the beef cattle that graze on pastures that were once rain forest. There is also the web of international debt that has grown between the industrialized and Third World nations which often forces those countries with rain forests to over-exploit them.

It is clear that we will not be able to maintain all of nature's wonders forever in pristine condition; that is the price we pay for development. Those areas that are an exceptionally rich biological resource need to be preserved intact as protected national parks, and the indigenous peoples should be given the chance to continue to manage their land without intrusions from outside. The rest may be sustainably managed in the way best suited to the land, whether that involves agriculture, forestry or industry.

Drowned rain forest, Amazonia.

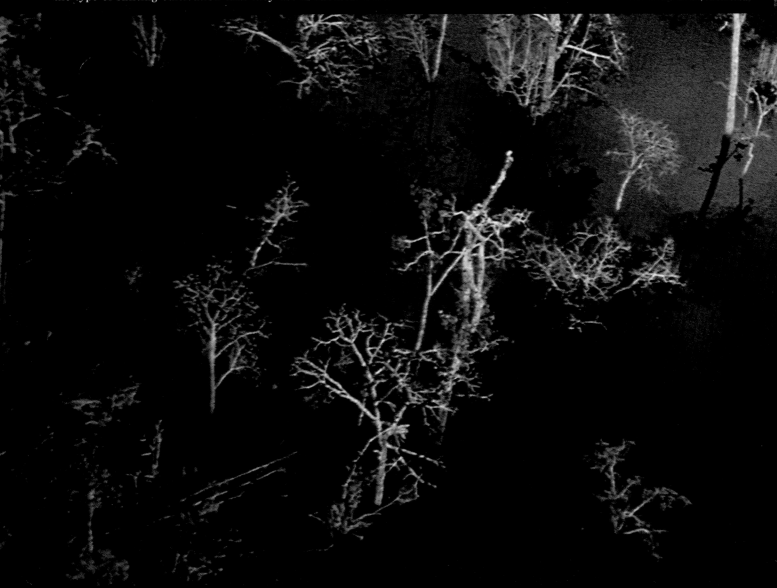

It is clear that we will not be able to maintain all of nature's wonders forever in pristine condition; that is the price we pay for development.

The impact of logging

Côte d'Ivoire, a major West African exporter of timber, is about to be completely logged out and expects to begin importing wood in the next two or three years. In Southeast Asia, timber production from the Philippines has declined through over-exploitation, a pattern now being repeated in parts of Malaysia. So far, the Amazon Basin has remained comparatively untouched.

Commercial logging is often seen as the major cause of deforestation in rain forests, but in fact it is almost never directly responsible for forest loss. Indirectly, however, it has more malign effects: migrant settlers frequently move into the forest along the loggers' roads and complete the deforestation illegally.

For tropical nations, timber is an important source of foreign currency. Today, the annual world trade in tropical timber exceeds US$8 billion in value. That comprises about 25 million cubic metres (880 million cubic feet) of raw logs, eight million cubic metres (280 million cubic feet) of sawn hardwood, and seven million cubic metres (250 million cubic feet) of plywood and veneers. Japan is currently the world's prime consumer, accounting for about 35 percent of the total tropical timber market, the rest of the Far East accounting for about 25 percent and Europe a further 13 percent. In the past, almost all of Japan's tropical timber has come from Southeast Asia, but now that production from the region is begining to wane, Japan is looking to Africa and Brazil as a future source of wood.

Commercial logging tends to be based on a selective system, so that only a small percentage of the trees is cut and removed from the forest. Of the thousands of species of tropical trees, only a relatively small number – including mahogany (*Swietenia macrophylla*) from Brazil, teak (*Tectona grandis*) from Southeast Asia, and okoumé (*Aucoumea klaineana*) from Central Africa – fetch a good price in the conservative international market. In spite of attempts to market a wider range of woods, probably only 50 species are widely exploited. Even if selective logging techniques are employed, damage to the forest can still be substantial because many of the nearby trees are brought down with the ones felled. Heavy machinery causes further damage to trees and compaction of the soil. But the amount of damage to the forest depends on how carefully the logging is done.

Who is to blame?

Recently, attitudes towards the logging industry have become polarized. The view taken by some Western non-government organizations, such as Friends of the Earth and the World Rainforest Movement, is that loggers are responsible for mass forest destruction, both directly through their logging, and indirectly by creating access for landless peasants.

Other groups such as the IUCN and the World Wide Fund for Nature (WWF) regard the timber trade itself as holding the key to saving the forests, despite its poor track record in the past. The point is that if the forests die, so does the lucrative tropical timber trade, which is not in the interest of the logging companies.

In the past, many commercial loggers have gone in and "mined" their concessions, taking out whatever they have wanted with little thought to the future. This is partly because logging concessions have been so short – only five or ten years in some countries. Tropical hardwoods grow slowly, so there is little incentive to bother with a long-term, sustainable system. If, on the other hand, these companies are given longer concessions (60 years is seen as the minimum) it will be in their interests to protect the forest. Using a trunk girth measurement, whereby only mature trees are cut, it should be possible to ensure sustainable production, with the harvesting of mature trees on 30- to 70-year rotations.

Critics of this idea point out that there is a negligible amount of truly sustainable forestry currently in operation. Although that may be true, it does not imply that sustainable management of tropical forests is not possible. Indeed, we should all hope that it is possible, because the only alternative if the tropical timber trade is to continue is to clear-fell forests and set up tree plantations. Ecologically, this is the worst possible scenario, because it reduces biological diversity to near zero.

Another tree falls (*above*) – Even though much of the work of removing cut trees from the rain forests to the timber mills is carried out with the use of large machines, the job of felling individual trees is usually done by men with chainsaws. The litter left behind in a logged forest can increase the risk of fire as was the case in Borneo in 1983 (see p. 168).

Moving the timber (*left*) – Out of the water, the trunks are cumbersome and extremely difficult to manoeuvre other than with powerful machines and tractors. Nonetheless, it is only out of the water that some types of wood can be properly assessed for processing.

Floating timber-yard (*left*) – Logs floated down to this timber-yard on Mindanao island in the Philippines are sorted out on the water before undergoing industrial processing in the plant behind. The pungent smell of waterlogged wood is redolent for kilometres around.

Shifting and shifted cultivators

Slash-and-burn: an emotive phrase that more often than not brings to mind pictures of destruction, of vast areas of smouldering tree stumps put to the torch as peasant farmers clear the forest for cultivation. But this technique has been used by generations of indigenous rain forest peoples throughout the tropics as part of a life-giving, sustainable forest agriculture system. Known as "shifting" or "swidden" agriculture (see page 95), this is often the only way in which the nutrient-poor rain forest soils can support crops.

Today, the problem is that as population pressure increases, the natural limitations of the shifting system are not respected. As the land is used more and more intensively, the fallow period becomes shortened, which leads to the over-working of the soil, decreasing soil fertility and reducing crop yields. In addition, the migrant settlers, who are not familiar with the forest, do not generally plant the wide variety of crops used by traditional shifting cultivators, which removes what is essentially a natural pest-control system. Planting monocultures makes the crop much more susceptible to pest infestations. Furthermore, the cleared plots are frequently much larger than those of the indigenous people, which means that forest regeneration during the fallow period takes significantly longer.

Without adequate protection from the elements, the soil is quickly degraded by the intense heat of the sun; the nutrients from the initial burning are soon washed away by heavy tropical storms, and the plots become infested with weeds and pests. After a few years, many settlers find that they are unable to support themselves; they then abandon their plots, or sell up to cattle ranchers (see pages 42–43) and move on down the road.

A combination of increasing population pressure and ignorance of suitable farming techniques has led to severe land degradation throughout the tropics. Vast areas of scrub and unproductive grassland – 30,000–40,000 square kilometres (11,500–15,500 square miles) in Papua New Guinea alone – bear witness to the unsuccessful attempts. In Laos, current estimates suggest that between 2,000 and 3,000 square kilometres (800 and 1,100 square miles) of forest are lost each year to the army of shifting cultivators. If present rates continue unchecked, all Laotian rain forest will have disappeared by the year 2030.

It is widely recognized that this sort of uncontrolled and unsustainable shifting agriculture represents the biggest threat to the future of the rain forests. In many countries landless peasants in areas of high population density are encouraged to move into less developed forest areas by governments. These people are thus essentially "shifted" cultivators.

Projects such as Sri Lanka's Mahaweli Regional Development Programme, Brazil's Transamazonia Highway Project, and Indonesia's transmigration programme (see box) have all been government-backed attempts to relocate people in response to population pressures. Their success has been extremely limited.

Rondônia, a Brazilian state the size of the United Kingdom that lies on the border with Bolivia, is one example of an area that has been badly affected by government colonization policy under the slogan "Land without people for people without land". In 1960, Rondônia had an indigenous population of about 10,000, who were traditional shifting cultivators. Following the paving of the BR364 road, the population mushroomed to more than 1.5 million by 1985; and between 1983 and 1985, 11 percent of the state's forest was felled. If the destruction continues, Rondônia will be almost completely deforested before the end of the century. In the neighbouring state of Acre, chainsaws have been handed out free as a vote-gaining exercise.

However, the majority of shifted cultivators are not part of government-backed projects, but move into forested areas spontaneously along roads built by governments, or by commercial logging or mining companies. These people frequently operate illegally, and are beyond the control of the government. But with few forest-protection controls, the destruction continues.

Existing roads (*main pic*) give shifting cultivators access to the rain forest (*inset*).

Transmigrant colony at Sorong, Irian Jaya.

Migration on a massive scale

The Indonesian Transmigration Programme is the world's largest programme for voluntary, government-sponsored migration. Since 1905 at least 2.5 million people have been moved from the crowded islands of Java, Madura and Bali to new settlements on the less densely populated areas in Sumatra, Kalimantan, Maluku and Irian Jaya (see page 168). It is estimated that perhaps as many as five million more people have moved without government assistance, but as a direct or indirect result of the programme.

The biggest drive to move people came at the beginning of the 1980s when more than 60,000 families were being moved each year at an estimated cost of US$10,000 per family. By 1987, the pace of transmigration slowed to around 10,000 families a year, as the programme ran out of money and as the Indonesian government became more sensitive to the problems transmigrants were experiencing when they were often settled on land that couldn't support them. Although transmigration has undoubtedly improved the lot of some families, many others have moved thousands of miles, yet ended up in city slums that have grown up in the outlying regions – notably in Irian Jaya. The Indonesian government has acknowledged mistakes, and hopes to continue the programme on a smaller scale, with more careful planning.

As a means of relieving population pressure on Java, the programme has been ineffective. Despite family planning measures, the population growth on that island still exceeds the number of transmigrants moving away. Other critics of the programme claim that it is nothing more than a means of "Javanising" the entire archipelago – something which has not been well received by the local Irianese in Irian Jaya.

Transmigration settlements have officially been viewed as centres of growth and development, but the more succesful sites have had to cope with a large influx of unassisted migrants who have been lured by the good reports of friends and families who were government-sponsored. The environmental consequences of the activities of these extra people have often been dire. In Lampung, Sumatra's southernmost province (the closest to Java), these migrants have degraded hillsides and forested land to such an extent that they have had to be relocated in neighbouring Bengkulu province. The integrity of the Barisan Selatan National Park between Lampung and Bengkulu is now being threatened by continued illegal forest clearance, and lax policing.

Cattle ranching

During the last 30 years the spread of beef cattle ranching has posed a serious threat to the rain forests in Latin America. The clear-felling of huge areas of forest has been given special tax advantages by governments in Central America and Brazil, and aid grants from the World Bank, to produce beef for domestic consumption and export to the North American and West European fast food markets. The combined herds of Nicaragua, Honduras, Guatemala and Costa Rica doubled to 9.5 million head of cattle between 1960 and 1980; during the same period a quarter of the forest in these countries was cleared, and the process is continuing even faster today.

Cattle ranching generally represents the third and final phase of forest degradation, logging and attempts at shifting cultivation by landless settlers. But even ranching is not usually sustainable for more than a decade, so the cattle men move on to new pastures as productivity falls.

The very low initial stocking rate of just one animal per hectare (2.5 acres) immediately after clearance is soon reduced as soil fertility declines. Five to ten years after clearance, each animal needs more than five hectares (12.5 acres). With such low stocking rates it is not surprising that meat productivity is not even one-tenth of that achieved by European farms. Some of the richer cattle men dose their pastures with fertilizers, but the end result is the same: the fertility of the soil declines and the pastures are invaded by weeds; under the trampling of hooves, the soil is compacted, exposed to the elements and then eroded away – in Costa Rica, it is estimated that for each kilogramme (2.2 pounds) of beef exported, 2.5 tonnes (2.2 US tons) of soil is lost.

A similar story of destruction has been unfolding in parts of Brazil's Amazonian forests. Although it represents the third stage of forest degradation, cattle ranching has often been the engine which drives the cycle. Many landless peasants deliberately clear forest with the intention of using it themselves for a couple of years and then selling it at a previously agreed price to a rancher. Sometimes the ranchers organize clearance of the forest, selling the best wood, but allowing peasant farmers to use the land for two years if they first clear the trees.

Many ranch owners are simply land speculators who are anticipating rising land prices following the opening up of Amazonia. Tax incentives and ranching subsidies ease the initial costs, while the sustainability of ranching is of little consequence because the real interest is in the value of the land. Legally, cattle ranching is viewed as "land improvement", and as such gives the rancher title to the land. Rich and powerful land speculators have been known to use fraud and violence to eliminate claims from smaller competitors such as peasant farmers, rubber tappers (see page 124) and indigenous peoples.

At last, the tax incentives that encouraged the ranching have been withdrawn. This should substantially slow the spread of ranching, in that without the incentives it is inherently uneconomic. Seeded pastures in many parts of the Amazon Basin lose fertility after just five years of ranching because the amount of phosphorus in the soil, which is an important nutrient for healthy pasture, quickly declines as soils are eroded.

At the state-run Brazilian Agriculture Research Agency in Belém, agronomists have been experimenting with new varieties of grass in the hope of transforming cattle ranches into profitable enterprises. Thus far they have had little success. However, Colombia's agricultural research institutions, together with the International Institute for Tropical Agriculture, Nigeria, have been experimenting with a combination of grasses and legumes. Once legumes have established themselves in a field, they begin to provide the nitrates which grasses require to flourish. Nitrates and organic matter from dead and decaying plants and animal droppings stimulate microbial activity in the soil. In this way a once-barren soil can be rehabilitated. This, perhaps, could mean that previously-exhausted lands could be reinstated as pasture land, thus ensuring that further forest is not encroached upon.

Cattle ranching in Amazonia leads all too soon to soil erosion.

Industry in the forest

To many tropical nations, their forests are not just a source of timber and land: they may conceal considerable mineral wealth beneath the trees, and the potential for damming rivers as a source of renewable hydroelectric power is often enormous. As a direct threat to rain forests, mining is a relatively minor cause of deforestation, although access roads, and the generally increased level of development in the region of mines, frequently attract landless settlers. The Amazon Basin certainly contains enormous mineral and oil wealth, as do parts of New Guinea, the Philippines and Indonesia.

Probably the largest and most ambitious mining project being developed in a rain forest is the Brazilian Grande Carajás Programme. Costing an estimated US$70 billion, it will cover an area in eastern Amazonia the size of France. At the core of the programme are the enormous deposits of iron ore which lie under the forest. At least 18 pig-iron smelting plants are being set up; the first, at Marabá in the state of Pará, started production in March 1988. These smelters will be fuelled with charcoal produced from virgin rain forest. When all 18 smelters are on stream, the charcoal they burn will consume a staggering 2,300 square kilometres (900 square miles) of virgin forest each year. Other commercial mining operations in Amazonia take a more responsible attitude. The land used for the Mineraçao Rio Norte mine is reafforested after the ore has been removed (see page 121).

Other industrial pressures on the world's rain forests come from illegal gold mining by masses of landless peasants-turned-gold-prospectors. Gold rushes in the southern Philippine island of Mindanao, and in various parts of Amazonia have resulted in the pollution of rivers with mercury (used in the separation of gold from the ore), and disruption to tribal peoples.

Flooding the forest

One feature of rain forests, as their name suggests, is that they have an extremely high annual rainfall, which feeds some of the world's great rivers including the Amazon, the Zaire (formerly the Congo), and the Orinoco. But harnessing this energy means building dams and flooding large areas of forest. Although many tropical rain forest habitats are adapted to seasonal flooding (see pages 16–19), the creation of a reservoir is usually a large-scale, extreme and permanent change.

The Amazon outstrips all its rivals in its hydroelectric capacity. It carries one-fifth of the Earth's entire fresh water supply through its channels every day. A conservative estimate puts the electricity that could be generated from the Amazon's thousand-odd tributaries at 100,000 megawatts. The Tucuruí dam was Brazil's first large hydroelectric project in Amazonia, flooding about 1,750 square kilometres (650 square miles) of rain forest. Under the Brazilian government's Plano 2010, some 136 new hydroelectric dams are planned. However, following the protest at Altamira by the Kayapó Indians (see page 112), the World Bank withdrew all its funding for Amazonian dams.

Apart from the sheer loss of forest, dams often cause major ecological problems. Silting is a major concern with many and may be exacerbated by deforestation of the watershed. For example, the Ambuklao dam in the Philippines has had its expected life reduced from 60 to only 32 years because silt is filling up the reservoir. In addition, when nutrients accumulate in a reservoir, they promote rapid growth of algae which upsets the ecological balance, and can result in the death of fish. The static waters of reservoirs additionally provide excellent breeding grounds for snails, mosquitos and other insects. Increased incidence of malaria and schistosomiasis is associated with dams. The Tucuruí dam is a case in point. The ecology of the river below the dam has been severely disrupted; fish and invertebrate species that were adapted to seasonal flooding of the Rio Tocantins are no longer able to breed; farmers are also affected, because their smallholdings were reliant on seasonal inputs of the fertile river silts. The impact on migrating river species such as the pink river dolphin (*Inia geoffrensis*) can only be guessed at.

Mineral wealth (*above*) – From the air it is possible to gain an impression of the extent of the Grande Carajás mine in Pará state, Brazil. Not only is it the site of the world's largest deposits of high-quality iron ore, but there are also large deposits near by of copper, gold, bauxite and manganese. Mining began in the mid-1970s and now the mining region is linked by a railway to the deep-sea port of São Luís on the Atlantic coast.

Scuba lumberjacks

When the Tucuruí dam was built, lack of developers' forethought meant that the trees in the valley, a valuable resource in themselves (estimated at 2.5 million cubic metres (90 million cubic feet) of prime timber), were not cleared before the reservoir was filled. Consequently, many of the flooded trees stand above the surface of the 20-metre (65-foot) deep reservoir, representing a hazard to shipping and fishing.

However, a new and unorthodox profession is now attracting the attention of local peasants and gold prospectors: underwater lumberjacking. Juárez Gómez, an ex-gold prospector, has recently invented an underwater pneumatic chainsaw, capable of working at depths of up to 50 metres (150 feet).

The costs of obtaining the wood are said to be only a quarter of those involved in normal logging, largely because it is relatively simple to bring the cut wood to the surface. High-quality wood such as mahogany can be removed for as little as US$5.50 per cubic metre (35 cubic feet) of wood. On the international market such wood may fetch up to US$900 per cubic metre.

Gold rush (*above left*) – During the early 1980s up to 50,000 mining workers and their overseers toiled daily at the Serra Pelada mine in Brazil. By 1986 more than 33,000 tonnes (29,000 US tons) of gold had been extracted . . . and what had once been a rolling hill had become a deep, flooded hole. Today, gold mining continues in hundreds of small-scale operations throughout the gold-rich areas of the Amazon Basin. There are also mechanized mining boats that search for gold in the Amazonian rivers.

Creating charcoal (*above*) – To extract iron from the ore in which it is mined requires extremely high temperatures. The most convenient way to produce such temperatures is to burn charcoal, itself produced from the wood of the rain forest. This is carried out in small, mud brick kilns by the *carvoeiros* (which literally means "carbon people"). The necessary skills are usually passed down within a family.

How rain forests work

The splendours of the tropical rain forests invariably make a profound impression on naturalists who visit them. Margaret Mee, the celebrated botanical artist, described a trip along the Rio Negro in 1967: "*Jara* palms grew in humid places along the banks, sometimes almost covered by the high water, their fibrous stems making wonderful homes for dozens of epiphytes. . . . As the sun touched the forest its last rays caught the red and blue plumes of macaws crossing the river in pairs or occasional threes, while oddly-beaked toucans, herons, noisy kingfishers and flights of parrots all on their homeward way kept my attention fixed. . . ."

More than a century earlier, another visitor to Brazil was equally impressed. "Here I first saw a tropical forest in all its sublime grandeur", wrote Charles Darwin in 1832, during the voyage of HMS *Beagle*, " – nothing but the reality can give any idea how wonderful, how magnificent the scene is. . . ." Darwin and his two eminent contemporaries, Alfred Russel Wallace and Henry Bates, were among the first European naturalists to explore the tropical rain forests, and the observations they made there helped to substantiate the theory of evolution that Darwin and Wallace both hit upon independently.

It is significant that the rain forests played such a role in shaping modern biology. Here is "the struggle for survival" at its most dramatic, evolution at its most profligate, adaptation at its most complex and intricate. The rain forest shows us life speeded up, intensified, portrayed in giant forms and vibrant colours. Even in the dense heart of the forest, where the gloomy shade cast by the canopy excludes most other plant life, there is still the endless bustling of termites and ants, audibly decomposing the forest, and the relentless upward push of the trees.

Darwin described how, following a rainstorm, an "extraordinary evaporation" took place. "At the height of a hundred feet the hills were buried in a dense white vapour, which rose like columns of smoke from the most thickly-wooded parts. . . ." Only recently have scientists begun to understand the significance of what he observed: three-quarters of the rain falling on to the forest is returned to the atmosphere by evaporation and transpiration (see pages 34–35). The water that rises as vapour eventually falls as rain again, but in the meantime it broods over the forest in the form of humid air and rain clouds, which moderate the heat of the sun and reflect some of its potentially damaging heat. The rain forest may create the moist air and clouds, but it is also true to say that *they* create the rain forest, by cushioning it from the full glare of the tropical sun. Science has revealed this fragile interdependence between climate and vegetation at the very moment that humankind seems intent on reducing the rain forests to shreds. The greatest fear of conservationists is that the rain forests can not survive once a certain proportion of the trees are destroyed, because the protective cloud cover would be lost.

Dissecting the forest into scattered fragments has other dangers too. Biologists think of these fragments as islands, surrounded by a "sea" of pasture or plantations. The analogy has proved useful in predicting what happens to such islands after they become cut off from other forests. Islands are invariably poor in species compared with the "mainland", although the exact reason for this is still debated. Studies of fragmented habitats around the world, and, more recently, in the Amazon Basin (see page 120), have confirmed that the "island effect" will occur. With each passing year, more species will die out, until each forest fragment reaches its true, and rather meagre, carrying capacity.

With the rain forest we would lose something infinitely precious. The effect on world climate is unpredictable, and possibly catastrophic. The effect on the biological richness of this planet would be devastating. The more we understand about how the rain forests work, the more we realize that they must be saved.

Flowering trees in the Amazonian rain forest.

Face signals – The bald uakari (*Cacajao calvus*) has a hairless face, whose flushed appearance may play a part in signalling to others of its species. There are two types, one with white fur, the other golden brown. Both are confined to the flooded forests of Amazonia.

Body artists – The Kayapó are shifting cultivators renowned for their tradition of body painting. Like most other native Amazonians, their existence is now threatened by gold mining and the destruction of the rain forest.

Dependent on water – Tree frogs are among the most distinctive forms of rain forest life. The air within the forest is heavy with moisture, allowing these thin-skinned animals to survive in the tree tops.

Evolution of the forests

The tropical rain forests have been evolving for almost 200 million years, beginning during the age of the dinosaurs, when a period of great drought finally came to an end and the planet enjoyed a warm, moist climate again. Forests grew up and covered most of the land, which was then still part of a single landmass or "supercontinent" known as Pangea.

Those ancient forests were very different from their modern counterparts because flowering plants – the group to which most present-day rain forest trees and plants belong – had yet to evolve. Instead there were gigantic conifers forming the canopy, with tree ferns, the palm-like cycads and the *Ginkgo* (relatives of the living maidenhair trees) making up the lower storeys.

Over the next 100 million years the flowering plants evolved from the ferns, and developed their curious relationship with pollinating insects (see page 72). Flowers were their most conspicuous development, but they were also much more adaptable plants, and this fitted them for life as opportunists in the forests, springing up in clearings where a mighty conifer had been thrown down by a storm. From these humble beginnings as pioneer species, the flowering plants developed larger and more robust forms that eventually took over the canopy as well.

The earlier plants have left behind some small reminders of their golden age, such as the ferns growing on the branches of trees as epiphytes. Larger survivors are few but they include the tree ferns, found in most tropical rain forests, and *Araucaria* conifers (relatives of the monkey puzzle tree of suburban gardens) which still flourish in the rain forests of New Guinea. Another curious survivor is the *Gnetum* vine, related to the conifers, and the only living member of this group to grow as a climber.

Continental drift

While the flowering plants were evolving, Pangea was slowly breaking up, and the continents moving apart to their present positions. The sea rose at times and inundated forest areas, eventually falling again and allowing the forest to regrow. Sometimes the sea cut one continent off from another, and both South America and Africa were islands for millions of years at a stretch. These events were of great significance in the evolution of the forest, and they help to explain the huge differences in flora and fauna between the rain forests of different continents.

It was while Pangea was disintegrating that the birds evolved (beginning about 150 million years ago) and the dinosaurs died out (65 million years ago) leaving the way clear for mammals to diversify and increase their size. Because these important developments occurred after the continents had separated, evolution took its own idiosyncratic course in each of them. The most striking differences are seen among the mammals, and this is well illustrated by the primates. The monkeys of the New and Old Worlds developed independently from a common but distant ancestor, and, while superficially similar (because they have adapted to similar ways of life), they show fundamental differences in anatomy. Apes developed from the Old World monkeys and are unknown in Central and South American rain forests, but large monkeys such as the howler monkeys have evolved to fill a similar niche. Madagascar, isolated from Africa before the monkeys and apes evolved, preserves an earlier phase of primate evolution in the form of the smaller-brained lemurs. Although eminently suitable for them, the rain forests of Australia and New Guinea have no monkeys or apes at all, their place being taken by marsupials such as tree kangaroos. The marsupials, although highly evolved, are descended from a very early stage of mammalian evolution. They only survived in the southern continent Gondwanaland (later to become Australia, Africa, South America and Antarctica) because placental mammals, which evolved later in the northern part of Pangea, were unable to reach Gondwanaland after it broke away.

In both Asia and the Americas, continental drift has enriched the flora and fauna of the rain forest, by mingling the species of two distinct regions. Australia and its associated islands finally

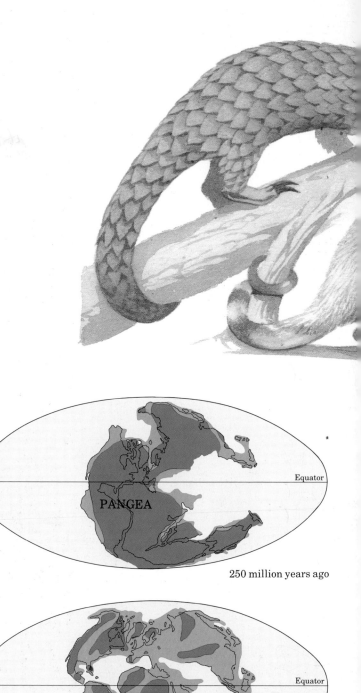

250 million years ago

100 million years ago

The birth of the forests – The continents rest on sections of the earth's crust which are in constant motion relative to each other, carrying the continents about with them. Around 250 million years ago, all the continents had been pushed together to form a single landmass, known as Pangea. This "supercontinent" was the birthplace of the present-day tropical rain forests. Before this, vast swampy forests covered much of the land, but these had dwindled and disappeared during a long period of drought. It was the end of this drought that signalled the birth of new forests. Most of the trees in the forests were conifers, cycads or tree ferns. The flowering plants, which are now dominant, did not appear until about 100 million years ago. With flowering plants came the great evolutionary expansion of the insects. Birds had already appeared by this

Tree pangolin

Tamandua

Convergent evolution (*left*) – Although the tamandua anteater (*Tamandua mexicana*) of Central and South America and the African tree pangolin (*Manis tricuspis*) look very similar – both are adapted to a life in the forest, living on arboreal ants and termites – they are not closely related, and have evolved independently on different continents. This is illustrated by the way in which each species protects itself from insect bites: the anteater has a thick hairy coat, whereas the pangolin is covered in scales, which also protect it from predators.

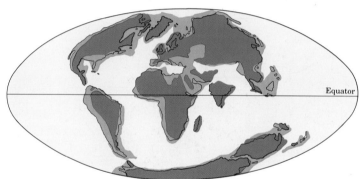

50 million years ago

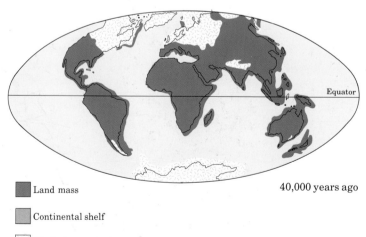

40,000 years ago

▪ Land mass

▪ Continental shelf

□ Glaciation

Positions of the present-day land masses are depicted by a black line on each map

Ancient forms (*above*) – Ferns, such as these giant ferns (*Alsophila armata*) in the Atlantic coast forest of Brazil, used to be far more common in the rain forest. This was before flowering plants diversified and became the dominant plant type. These ferns are comparatively intolerant of shade and prefer a wet environment.

Tree ferns (*above*) – The Malaysian tree fern (*Cyathea contaminans*) can reach a height of 15 m (50 ft), and is one of about 700 living tree fern species, many of which have a very limited distribution.

time, but were still relative newcomers, while mammals had yet to appear. As this was happening, Pangea was breaking up, carrying sections of rain forest away to continue evolving in varying degrees of isolation. Throughout this time, despite climatic variations, there was always some forest cover in at least part of the tropics. This continuum is an important factor in the evolution of the rain forests and the complexity of relationships found in the forests today.

separated from Gondwanaland and drifted northwards, to collide with the continental mass of Southeast Asia about 15 million years ago. Successful species from each continent migrated into the new lands now available to them. Eucalypts, cockatoos and marsupial phalangers spread northwestwards from Australasia, while rattans, woodpeckers, mice and many other species spread southeastwards from Asia.

Falling sea levels finally reunited South America with North America less than five million years ago, allowing an exchange of animals between these continents. Among the existing fauna of Amazonia are relicts of the continent's island past, such as the anteater, armadillo, sloth, agouti and capybara, mingled with newcomers from the north, such as the peccary, squirrel, jaguar and tapir.

The Ice Ages
Throughout these epochs, the climate fluctuated many times, becoming hotter or cooler, wetter or drier. But the most dramatic changes came quite recently, with the series of Ice Ages that occurred between 2.5 million and 10,000 years ago. Each time the ice caps crept towards the equator, the tropical regions would have become both cooler and drier, so that some areas of rain forest became seasonal forest or savanna. No one knows how large an area would have been affected, but it is believed that the rain forests of America and Africa were more depleted than those of Southeast Asia.

The effects of the ice ages on the tropical rain forests is currently a matter of debate. During the past 20 years, the idea that the Amazonian rain forest was reduced to a number of small islands or "refugia" has gained widespread support among scientists. These supposed refugia, which correspond to areas containing an unusually large number of species today, were said to have acted as "Noah's arks" during the Ice Ages. It has also been argued that the splintering of the forest into isolated refugia encouraged the evolution of a great many new species – although the opposite argument, that the Ice Ages reduced species diversity by causing extinctions, has sometimes been put forward, and is equally plausible. Recently, many people have begun to question the evidence for refugia. As different animal groups have different refugia, it is suggested that such areas owe the predominance of their life forms to some other underlying cause, such as a favourable local climate or soil.

The rain forests today
Even outside such centres of diversity, modern rain forests are remarkably rich in species. A hectare (2.5 acres) of Malaysian rain forest can contain as many as 180 different species of tree, whereas a temperate wood would be unlikely to have more than ten. A handful of these 180 species would be reasonably common, the rest being extremely rare, perhaps only one or two individuals per hectare. The same richness and variety is seen among smaller plants and plant-eating animals, especially insects, but there is far less diversity among the predators. These generalizations hold good for all continents, but South American forests are the richest in species, and those in Africa notably poorer than the South American or Southeast Asian rain forests. Within each continent, some types of forest are more uniform than others, notably mangrove swamps, which never boast more than 25 different species of tree.

Why tropical rain forests should be so rich in species is a more difficult question than it appears. It seems certain that there is a variety of factors at work, and the mix of reasons may be different for different forests.

One common factor is the high input of energy from the sun (see page 61), creating bountiful conditions for growth, combined with a shortage of soil nutrients. The nutrients tend to be distributed patchily, and this encourages different species to evolve, capable of dealing with the infertile conditions in varied ways, or exploiting different patches of soil.

Chemical warfare – The colour of these lycaenid caterpillars (*Eumaes mynas*) advertises the fact that they are highly toxic, but the poisons that they contain are not their own. To deter leaf eaters, the plant species on which they feed is poisonous; but the caterpillars are able to incorporate this poison into their bodies without any harmful effects. The plant's defences are not as weak as they may at first appear, in that it has only one other insect predator. Such relationships are typical of the rain forest where there is intense competition.

In place of monkeys – The rain forests of New Guinea and Australia contain no monkeys or apes. Instead there are marsupial tree kangaroos, such as this Matschie's tree kangaroo (*Dendrolagus matschiei*) in the rain forest of Papua New Guinea. These kangaroos are not particularly well adapted to an arboreal lifestyle. Instead of making their escape through the trees when threatened, they tend to jump to the ground and then hop away.

Regenerating rain forest: Nature's own experiment

When the volcanic island of Krakatau erupted in 1883 it left behind four small islands, the fragments of its former self, devoid of life and covered by a layer of ash up to 100 metres (330 feet) deep. Naturalists then working in Indonesia realized the value of this "natural experiment" in revealing how a tropical rain forest would develop, starting from scratch. Regular surveys of the islands were made over the next 50 years, and the study programme was resumed more recently, in 1979.

The volcanic ash of Krakatau was first colonized by a slimy film of cyanobacteria, then by grasses and ferns. Small bushes followed, and within ten years of the eruption some tree saplings had become established. It took only 40 years for the islands to regrow a mantle of trees, but this was just a semblance of the true tropical rain forest. Closer inspection showed it to have only 36 species of tree, and even by 1979, when the rain forest stood 30 metres (100 feet) tall in places, there were only 60 species of tree in the inland areas. Almost all of these were trees typical of "secondary forest" – the pioneer species that spring up in light gaps (see page 66). A few of the trees that characterize mature rain forest had arrived but were extremely rare. This lack of plant diversity was reflected in the animal life, with only five species of butterfly in the interior forests.

Krakatau is only 40 kilometres (25 miles) from the coasts of Java and Sumatra, and relatively easily colonized. Of the plant species established by 1934, 41 percent were wind-dispersed and 28 percent animal-dispersed, the rest being carried by water. Islands farther from the mainland must rely solely on animal-dispersed seeds. Jarak, 64 kilometres (40 miles) from the coast of Malaysia, was smothered in volcanic ash about 34,000 years ago. A survey found only 93 species of flowering plant, a paltry total, of which only two were brought there by the wind. São Tomé, a volcanic island in the Gulf of Guinea, 300 kilometres (188 miles) from Africa, is almost entirely populated by animal- and water-dispersed plants. On all these islands, figs are conspicuously successful, a testimony to their prolific fruiting and lavish seed production.

A recent study on Krakatau underlines the importance of animal dispersers. This study found that once an island has a certain number of fruiting trees, it begins to attract seed dispersers such as birds and bats in far greater numbers. These bring in fresh seed from the mainland and enrich the species diversity of the island, leading to a greater year-round supply of fruit. This self-perpetuating cycle of fruit and fruit eaters should eventually lead to a far richer rain forest. But the process is painfully slow, because gaps in the canopy open so rarely, and the stock of seeds in the soil is so poor compared with established forest. The chances of a canopy-tree seed being in the right place at the right moment are, as yet, extremely small. Biologists working on Krakatau believe that it will take several hundreds, possibly thousands, of years for the forest to regain its former glory.

(*Inset*) Having been smothered in ash from Krakatau, the island of Rakata is once again covered in vegetation.

(*Main pic*) Regenerated forest can be seen around the shores of Anak Krakatau, while the volcano in the centre still smoulders.

Another factor, stemming from the high solar input, is the great height of the forest trees, which creates a massive three-dimensional structure (see page 62). Within this framework, a great many smaller plants, including climbers and epiphytes, can lodge and grow. The variety of these plants creates a tempting range of food sources and hiding places for small animals.

Climate is also a common factor, and this may be crucial. In other forests there is a winter or a dry season that interrupts the life cycles of insects and usually reduces their numbers. Without this "environmental sieve", the rain forest insects find survival relatively easy, and they have been free to diversify.

The unfettered proliferation of insects has undoubtedly exerted pressure on trees and other plants, because many insects feed on their leaves or seeds. The massive chemical armoury of most rain forest plants is a response to the pest problem. Some biologists believe that this "pest pressure" has also contributed to the diversity of plants in the rain forest. In developing chemical defences, a plant species inevitably encourages the evolution of insects that can overcome those defences. These insects evolve into specialist herbivores for that plant. (The effect of highly selective herbivores is well illustrated by the growing of plantation trees, such as rubber, in distant parts of the world. Initially, such trees do far better than in their native land, thanks to the lack of specialized insects that can attack them. But within a few years, the insects themselves are inadvertently introduced, usually with catastrophic effects.) If a tree or other plant is

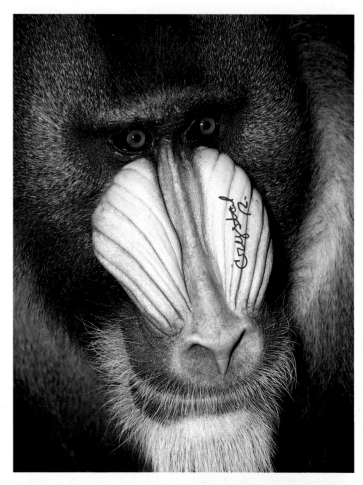

> **In developing chemical defences, a plant species inevitably encourages the evolution of insects that can overcome those defences. These insects evolve into specialist herbivores for that plant.**

immune to most insect herbivores but highly susceptible to a few, it is most vulnerable when growing near others of its kind. The specialized insects can move from one tree to the next with ease, resulting in a local population explosion. This has been observed where trees grow in single-species clumps in the rain forest – they are susceptible to massive pest attacks, usually by leaf-eating caterpillars, and can even be killed by them. Attacks of this type probably prevent any one species from becoming too common, and thus favour diversity.

A final factor in the diversity of rain forests, as in any forest, is the degree of disturbance to the canopy. If there are fairly frequent tree-falls, but the damage is not too extensive, then pioneer species, smaller trees, climbers and epiphytes will flourish (see page 66), greatly increasing the species diversity. By comparison, a rain forest that remains unchanged from one decade to the next will be poorer in species, as will a forest that suffers frequent large-scale degradation.

It happens that many tropical rain forests are subjected to the sort of regular, small-scale disturbance that promotes diversity. In the hilly terrain of Southeast Asia, and the montane forests of the Andes, landslides are a major factor. In central Amazonia, erosion of the land by rivers, which are constantly shifting their courses across the Amazon Basin, creates disturbed areas in the forest. On forested islands, such as those of the Caribbean and Indonesia, cyclones and hurricanes periodically flatten tracts of forest, or fell the tallest trees. All these natural cataclysms are only temporarily destructive, resulting ultimately in a richer and more diverse forest.

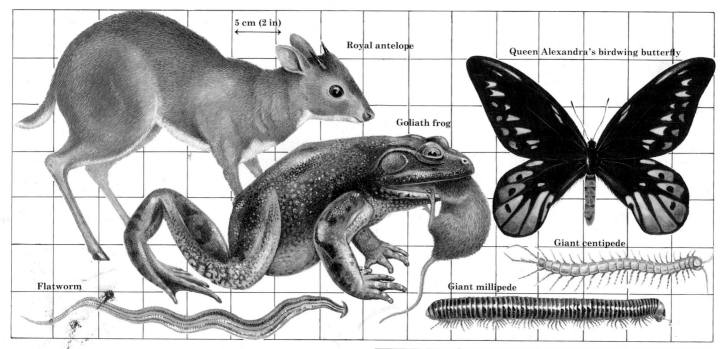

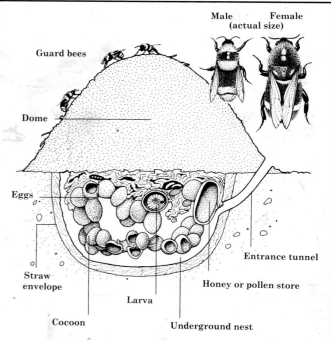

Getting the message across

(*left*) – The difficulty of communicating in the rain forest is reflected in the extremes to which certain animals have gone to get their message across. The mandrill (*Papio sphinx*) lives on the forest floor, where its bright face-mask can easily be seen through the undergrowth. But this colourful sign has its price. In older males the pale blue tissues on each side of the muzzle become so swollen that they obstruct the animal's vision.

Morning chorus

(*left*) – At dawn, South American rain forests reverberate with the calls of the black howler monkey (*Alouatta caraya*), as each group re-establishes its territory for the day. Compared to the mandrill, the coloration of its face is dull, reflecting this reliance on audible signals.

Extremes of size

(*above*) – The constantly warm climate of the rain forest has allowed some cold-blooded invertebrates to become giants, while life in the dense undergrowth has led to the evolution of dwarf mammals. The royal antelope (*Neotragus pygmaeus*) is the smallest antelope in the world, standing only 30 cm (12 in) at the shoulder. In contrast, the Goliath frog (*Conrana goliath*), which lives in deep forest pools, can grow to a length of 80 cm (32 in) and weigh up to 15 kg (7 lb). On and near the forest floor live giant millipedes (*Doratogonus* spp.), which feed on dead plant material, predatory giant centipedes, such as *Scolopendra subspinipes*, and giant snails and the flatworm *Bipalium kewense*. The largest butterfly in the world is also found in the rain forest. The female Queen Alexandra's birdwing butterfly (*Ornithoptera alexandrae*) has a wingspan of 28 cm (11 in).

The aggressive bumble bee

In temperate regions, most species of bumble bee nest underground, using the abandoned burrow of a mouse or vole. There is no sign of the nest above ground, and the entrance is not guarded – indeed, bumble bees rarely sting unless greatly provoked. The behaviour of bumble bees in the tropical rain forests is in striking contrast. Most nest in trees, and only one species (*Bombus transversalis*), from the Amazon Basin, makes underground nests. The bees of this species build a large dome of leaves, twigs and plant stems over their nest, something that no other bumble bee does. Up to five bees patrol the heap constantly, and make frenzied attacks on any animal that approaches. Their aggressive behaviour is typical of the rain forest, where competition for resources is particularly fierce and predation more intense.

Evolving together

Close and mutually beneficial relationships between two species – often described as mutualism or symbiosis – are known the world over. But the tropical rain forests are peculiarly rich in such relationships, and they have spawned some of breathtaking complexity, such as the bizarre pollination mechanism of the figs or the delaying tactics employed by *Amorphophallus variabilis* (see pages 72–73).

The absence of a winter or dry season may be partly responsible for this, because many such relationships involve insects, and an insect that is abundant all year round is a far more reliable partner. In the case of plants, the great distance between individuals of the same species creates a need for effective and energetic pollinators, while the competitive nature of the rain forest environment demands good seed dispersal. So there have been strong incentives for plants to form close ties with animals. Added to this, the rain forests have generally enjoyed a relatively long period of climatic stability in which coevolution – the process whereby two species adapt to each other – could proceed unhindered.

An intriguing feature of the rain forests are the extensive networks of relationship that have developed, sometimes involving a dozen or more species. For example, a water-storing bromeliad has relationships with its pollinators, its seed-dispersers, the tree on which it lives (see page 68), the many animals that live or breed in its water stores, and those that live in the debris around its roots. Some of these will be give-and-take relationships; others more one-way. Often there is one central relationship upon which the others depend, such as that between ants and ant-plants or between figs and fig-wasps. Such associations have been described as "keystone mutualisms". The great concern of all conservationists is that, as the rain forests become increasingly fragmented, such interdependence will crumble for lack of a keystone animal or plant.

A useful parasite (*left*) – The cuckoo-like giant cowbird (*Scaphidura oryzivora*) of Latin America, is a "brood parasite" which lays its eggs in the hanging nests of caciques and oropendolas. Unlike the cuckoo, the giant cowbird chick shares the nest with the hosts' chicks, but it is nonetheless parasitic, because it deprives them of much of their food. However, the cowbird chick does its nestmates some good, because it eats some of the parasitic botflies that enter the nest, and pecks off the botfly maggots from the other chick's flesh, thus reducing their level of infestation. When caciques and oropendolas nest close to the nests of aggressive bees or wasps, the birds chase away giant cowbirds, but when they are nesting alone they tolerate the cowbirds, presumably because the good they do outweighs the bad.

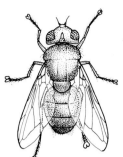

Wasp guardians (*left, above*) –
Oropendolas and caciques
provide a good example of the
complex webs of
interrelationships that are
typical of the rain forest. They
build hanging nests (*left*) to keep
out predators, but are
nonetheless plagued by parasitic
botflies (*above*). These lay their
eggs on the nestlings, and the
maggots eat into the young birds'
flesh. As a defence, the birds try
to nest in a tree where there are
stinging wasps or biting bees,
both of which are highly
aggressive in defence of the area
around their nests. The wasps or
bees reduce the number of botflies in the vicinity, as well as
deterring egg- and nestling-
predators such as opossums,
toucans and snakes. These
aggressive insects also attack
their avian neighbours at first,
but these attacks soon decline,
and the distinctive musty smell of
the birds is believed to play a
part in familiarizing the wasps or
bees with their presence. A third
element in this network of
relationships is the presence of
non-aggressive wasps and bees.
They build their nests near by
and, like the birds, benefit from
the furious self-defence of their
insect neighbours, although
giving nothing in return.

The black palm and the agouti

Coevolution – the process by which two species adapt to
each other – is a complex business, often involving many
other species besides the two main players, and rarely
resulting in a perfect relationship. Few associations
illustrate this as well as the unlikely alliance of the agouti
and the black palm (*Astrocaryum standleyanum*.)

The black palm guards its unripe fruit with needle-like
spines on the trunk and leaves. Only when ripe, with the
seed protected by a rock-hard case, do the fruits drop to the
ground. Once there, the rich fruity smell attracts pacas,
coatis and opossums, which eat the fruit and discard the
seed. Although this disperses the seeds, it does them little
good. Peccaries find them and crush the seed cases open to
get at the seeds, while spiny pocket mice and squirrels
gnaw their way through. Those seeds that escape fall
victim to bruchid beetles which lay their eggs on them; the
larvae can bore through the hard seed coat.

The only animal that actually benefits the black palm is,
paradoxically, an accomplished seed thief, the agouti. It
eats some of the seeds, but it buries many more for later
use. Before doing so, the agouti instinctively strips away
the fruit from the case. Experiments have shown that the
seeds will only survive when treated in this way. If buried
unpeeled, they fall victim to bruchid beetles, which have
already laid their eggs on the skin of the fruit, and whose
larvae burrow through to the seed after burial.

The puzzling aspect of this relationship is that the fruit,
which must have originally evolved to tempt seed dispersers, is usually discarded by the agouti. It seems likely
that when the tree's dispersal strategy first evolved,
bruchid beetles, and possibly mice, were less common. So
seeds that were left on the surface, but stripped of their
fruit, had a reasonable chance of germinating. The
increased numbers of beetle larvae and small rodents
mean that the agouti is now the key to the survival of the
black palm.

Feeding the forest

Seeing the luxuriant growth of the tropical rain forests in Southeast Asia, the early European colonists eagerly imagined the abundant harvests that could be obtained from farms and plantations. The forests were cleared and crops planted, but the results were largely disappointing – after a few years, the land produced little. In rain forests elsewhere in the world it was the same story, and the tragedy is still being repeated today as landless peasants move into the rain forests of Amazonia and Indonesia. Once the forest has gone, the soil deteriorates, usually leading to the abandonment of the land within a few years.

The standard explanation for this is that the soil is lacking in the mineral nutrients needed for plant growth – principally nitrogen, phosphorus, potassium, calcium and magnesium. According to this view, the forest trees themselves store most of the nutrients in this ecosystem, and are adapted to scavenge any nutrients in the fallen leaves and other forest litter. Nutrients are quickly sucked back into the living components of the ecosystem, and few are left in the soil to be washed away by rain.

There is one type of rain forest that fits this description particularly well, and that is the lowland rain forest of the Amazon Basin, although even here there is enormous variation. The tall *terra firme* forests which grow on acidic soils are short of calcium, and in some parts no snails are found, there being too little calcium to build shells. The heath forests, known locally as *caatinga*, which grow on coarse sandy soil, have enough calcium but are lacking in nitrogen and phosphorus, which stunts the growth of the trees.

The underlying problem in the Amazon Basin is that the soil and subsoil are naturally infertile and have been further drained of nutrients by the incessant rain that has fallen over many millions of years. In temperate forests, nutrients are leached away, but the weathering of the subsoil, which produces fresh soil, makes good the loss. In the Amazon Basin, the subsoil itself is already impoverished.

To make matters worse, the Amazonian soils are lacking in substances which can hold nutrients. This is particularly true of the sandy soils underlying the *caatinga*, and any nutrients washed into the soil from the leaf litter are quickly leached away. The trees are adapted to cope with this situation by having a spongy mat of tangled rootlets on top of the soil to intercept nutrients before they are washed away. From these root mats, vertical rootlets grow upwards into the leaf litter to grip fallen leaves or fruit in a hungry embrace. The roots are coated by fungi, known as mycorrhizals, which attach themselves to decomposing leaves, channelling nutrients such as phosphorus directly back to the root. Such partnerships between plant roots and fungi are known in other parts of the world, but are of particular importance in the rain forest.

In the interests of nutrition, the trees on these depleted soils are prepared to sprout roots anywhere. Hollow trees may produce internal rootlets to tap the layer of debris left by animal inhabitants. Trees also send rootlets into the debris that collects around epiphytes on their branches, to scour this for minerals. The epiphytes themselves have various devices for capturing extra nutrients (see page 66–67): the pools of water that collect in bromeliads supply the plants with minerals from fallen leaves and animals living in them.

Compared with these starvation conditions, some rain forest trees live a life of luxury, particularly those growing on volcanic soils. None of the forests studied in Asia, Africa or Australasia is so impoverished that the nutrients are all locked up in the vegetation. Even within the Amazon Basin there are areas of more fertile soil, especially in the flooded *várzeas* along the banks of the Amazon and its tributaries. Yet on a world scale, all tropical rain forest soils are somewhat infertile, due to the perennial leaching by rainfall. The trees also have to contend with the effect of rainfall on their leaves, washing out nutrients such as potassium which must then be recouped from the soil by the roots.

Swallowtail butterflies: the attraction of salt
These butterflies have congregated to feed on a river bank in Sarawak, Southeast Asia. The cause of all the interest is a puddle of urine left by a passing mammal, which contains salts the butterflies lack. Usually it is only the male butterflies that come to feed: they provide salts along with their sperm when they mate, and the female uses the salts to replenish those lost during egg production. It seems likely that the females prefer those males which can supply salts. While feeding on urine, the butterflies take in a great deal of liquid, extract the salts, and eliminate the rest – every few seconds, a large drop of water is squirted out of the anus. Butterflies also feed at river margins and on dry, salt-rich ground. Some species are attracted to dung or to animal corpses, possibly in search of nitrogen compounds or other nutrients.

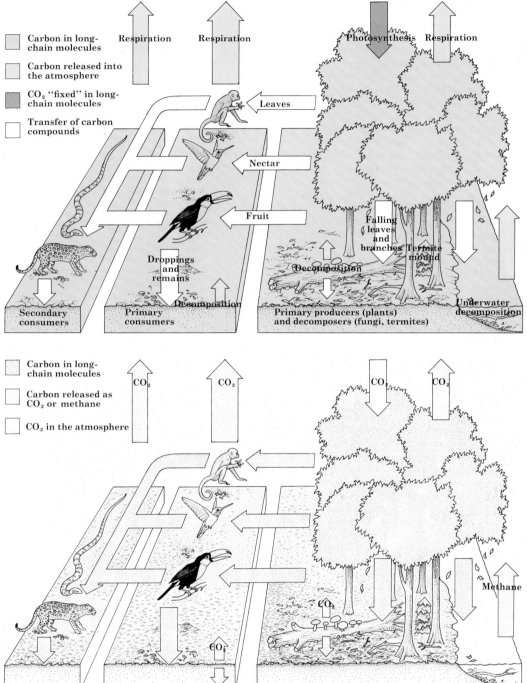

Forest fungi (*above*) – The fruiting bodies of this bracket fungus are emerging from a piece of dead wood. Not all fungi are decomposers. Some are parasites, feeding on live wood.

Using the sun's energy (*right*) – Life on earth is based on the power of the sun's rays. The same radiant energy that can blister human skin or melt a glacier is harnessed by plants and used to make sugar. In a complex chemical reaction, known as photosynthesis, the plant combines carbon dioxide (CO_2), which its leaves extract from the air, with water (H_2O) drawn up from the soil by its roots. Six atoms of carbon are linked together in a chain to form a molecule of glucose ($C_6H_{12}O_6$) and surplus oxygen is returned to the atmosphere. Carbon chains such as this are the basis of all life's diverse chemistry. Although often under cloudy skies, tropical rain forests receive ample radiation from the sun, simply because they are near the equator. When combined with high temperatures and plentiful supplies of water, this fuels an expansive growth of trees and other plants. The productivity of a tropical rain forest is colossal. Each year, it creates about 25–30 tonnes per hectare (10–12 tons per acre) of new growth, twice as much as a temperate oak wood, and more than three times as much as a northern coniferous forest. A large rain forest tree can make more than 1.5 kgs (3 lbs 5 oz) of pure glucose a day. Of this, 60% is broken down again as the tree respires, releasing energy to fuel the plant's life processes. The other 40% is transformed chemically into substances such as proteins or cellulose, to make the fabric of the tree. Plant-feeding animals absorb these molecules, alter them to suit their own needs and use them to build up their bodies.

Carnivores do the same with the animal bodies they consume. At each stage in this food chain the complex carbon molecules the plant originally built up are passed on. About 10% are rearranged chemically, but not broken down. The rest are split apart, as the plant or animal respires, to unleash the energy they contain. Those complex carbon molecules that remain, in fallen leaves and trees, animal dung, or the dead bodies of animals, provide food for fungi, bacteria and other decomposers, which break them down. With this last step, the energy originally invested by the sun is dissipated, and the remaining carbon chains dismantled. Much of the carbon returns to the air as carbon dioxide. But where decomposition takes place under water, as happens in swamp forests and the seasonally flooded *várzeas* and *igapós* of the Amazon, a different process occurs. There is no oxygen to make carbon dioxide, and methane (CH_4) is released instead. The same breakdown process occurs in the intestines of termites, which are major rain forest decomposers, consuming up to 17% of the leaf litter in some Malaysian forests. Methane is an important greenhouse gas, but it is not known whether tropical rain forests are net producers or net consumers of this gas.

In some of the Asian forests there is a suspicion that not all the nutrients present in the soil are accessible to plant roots, because the trees show adaptations for scavenging nutrients above ground, similar to those of Amazonia. Heath forests and forests growing on the crests of ridges in Sarawak both have root mats, and some Asian rain forest trees support colonies of bacteria in their leaves which can make usable nitrogen compounds from the nitrogen in the air. (Although "nitrogen-fixing" bacteria that live in plant roots are common around the world, only these rain forest trees have such bacteria in their leaves.)

The heath forests are also home to many carnivorous plants, which capture insects for extra minerals, and to the ant plants (*Myrmecodia* spp.) whose tiny inhabitants provide debris to nourish them. The soils of these heath forests cannot support agriculture, as shown by the Indonesian transmigration programme to central Kalimantan (see page 41), nor can they be reafforested if the soil has been at all disturbed. The exact reasons for this are unknown, but it seems that without the forest cover irreversible changes occur in the soil.

Decomposers: feeding on rotting matter

The recycling of nutrients through the breakdown of dead animals and plants is important in all ecosystems, but particularly so in the rain forest where low soil fertility often leaves little to spare. The organisms that act as decomposers are known as "saprotrophs" which literally means "putrid nourishment". They have become adapted to feed on dead material – fallen leaves and branches, animal dung and urine, corpses, moulted skin and any other remains. In feeding on these substances, they also break them down.

Often one group of decomposers – usually insects – begins the process by chewing into the remains, opening them up to invasion by fungi and bacteria. These can tackle a wider range of chemical compounds than can insects and thus take the process of decomposition further.

One of the most difficult tasks chemically is breaking down cellulose, the major constituent of leaves. Fungi are able to do so, using special enzymes, and so can some bacteria. Even more challenging than cellulose is lignin, the fibrous element in wood, and the chemical equivalent of a burglar-proof safe. A few fungi can tackle lignin and are thus of major importance in any forest. So too are the termites, which can account for as much as 70 percent (by weight) of the invertebrates in the leaf litter. Some break down plant material with the help of protozoa (single-celled animals) or bacteria living in their digestive system. Others enlist the help of fungi to breakdown partially digested leaves.

It is important to remember that decomposers feed on the remains of living things because it suits them – the natural world's "rubbish" represents an abundant and accessible source of food. Their action also releases nutrients such as nitrogen, phosphorus and iron from dead remains, which is of enormous benefit to other animals and plants. Without decomposers life would eventually grind to a halt for lack of nutrients. But their assistance to other life forms is just an unintentional side-effect: decomposers are no more altruistic than other living things.

Decomposition is rapid in the tropical rain forest, thanks to the great warmth and moisture in the environment, but it is not as exceptionally fast as is often believed. The rate at which the leaf litter breaks down is equalled by some temperate broadleaved forests. Indeed, there is one type of tropical rain forest where the rate of breakdown is actually very slow. In montane forest, also known as cloud forest, the ground is waterlogged, because the forest is usually shrouded in mist, making the atmosphere exceptionally moist. Temperatures are fairly low, owing to the altitude, and the combination of cold and waterlogging (which limits the amount of oxygen available) puts a brake on the decay processes. A layer of peat – unrotted acidic plant material – builds up on the forest floor and nutrients are imprisoned in the peat layer, so that the trees grow up stunted and gnarled.

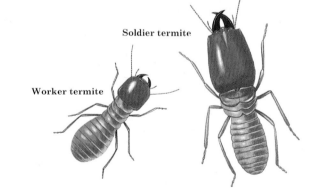

Soldier termite

Worker termite

Termites (*top, above*) – Some termites' nests are built largely of their droppings, glued together with saliva to set like mortar. Those termites that cultivate fungus gardens, however, build nests out of soil, because their droppings feed the fungi. Nests are often rich in mineral salts, and may be eaten by monkeys and other animals, to supplement their diet or to aid digestion of leaves. Unoccupied nests are a good seed-bed for sprouting trees.

Catching the nutrients (*right*) – With the leaf litter kicked away by passing feet, the tangled mass of tree roots, along with the mat of fine rootlets that overlies the soil, is revealed on this trail through the Sarawak forest. Such root mats are typical of rain forests on extremely infertile soil. They absorb minerals that are leached out of the decomposing leaf litter and prevent them from being washed away.

Structure by strata

The most lasting impression of a rain forest is one of overwhelming, cavernous greenery in which there is little evidence of other life. With year-round warmth and ample moisture, plants rule in the rain forest as nowhere else on earth. Only 0.0002 percent of the biomass (dry weight of living organisms) in one area of Amazon rain forest was found to consist of animals, and almost 70 percent of these were decomposers (see pages 58–59).

As well as being the warmest and wettest habitat for plant life, the interior of the rain forest is also one of the darkest. Of all the regions of the world, the tropics receive the most sunlight, but only one to two percent of this reaches the forest floor. As a result, competition for light is the driving force behind the structure of the rain forest. Plants tend to form bands or strata of foliage at different levels above the ground, each one filtering out yet more light, which successively reduces the amount of light and the temperature, while increasing the humidity, of the level below. Each level therefore has its own microclimate.

A walk through any mixed temperate forest will also reveal strata: a canopy of the tallest trees, with an understorey of smaller or younger trees, beneath which grows a layer of shrubs. At ground level a layer of herbaceous plants grows among the seedlings of shrubs and trees. What is unique in the rain forest is the range of different strata, the vast numbers of different plants in each layer, and the enormous difference in light levels between the upper and lower strata.

More than 40 types of rain forest have been described by botanists. Factors such as soil, altitude and latitude (both of which cause variations in temperature and rainfall), and whether or not the forest is subject to flooding, hurricanes and other natural catastrophes, and to what degree, influence the numbers and distribution of species and the overall structure. Some lowland rain forests may have five strata, montane forests two or three, and certain types of forest are arguably not stratified at all. In any case, the strata are often obscured by the sheer amount of vegetation, especially that of the climbers and lianas (see pages 68–69) which link all the levels. Although some plants, such as epiphytes (see pages 66–67), complete their life cycles at one particular level, others are able to grow towards the light from the shaded forest floor. Some canopy trees have large seeds, which fuel an initial spurt of growth; but when established, the saplings can endure many years of minimal growth, waiting for a light gap to appear. Although shade-tolerant when young, most of these trees need full light for mature growth, flowering and fruiting. Other canopy trees can germinate only where there is plenty of light.

The densest layer of vegetation in most primary forest is at about 20–30 metres (65–100 feet) above the floor, where the canopy trees, laden with epiphytes and climbers, branch into countless umbrellas of leaves. This zone is exposed to the full glare of the sun, bathed in temperatures of 32°C (90°F) or more, although the humidity is only about 60 percent. It is the powerhouse of the forest, where most of the photosynthesis (see page 57) takes place. Much of the flowering and fruiting also takes place in this zone, and attracts many insects and larger animals that consume the produce.

At the other extreme is the forest floor: the air is still, the humidity around 90 percent, and the temperature around 28°C (82°F). In the unrelieved gloom, among the buttress and prop roots of the trees, only shade-tolerant plants can survive. At ground level, some primary evergreen rain forests are so dark that the forest floor may be fairly bare and easy to walk through; but in more open forests, where the light levels are higher, the forest floor is a "jungle" of herbaceous plants and juvenile trees, shrubs and lianas. A more luxuriant understorey is also characteristic of the monsoon forests of Asia, which are found at the edge of the true rain forest in Indonesia, Thailand and India. This type of forest has a marked dry season in which most of the trees lose their leaves, enabling many more plants to thrive on the forest floor.

Drip tips (*above*) – Many of the plants on the forest floor have narrow, downward-pointing tips. These "drip tips" speed the loss of water from the leaves, an important adaptation for plants growing in heavily shaded, wet conditions. Without these tips, the leaves would be almost constantly wet. This surface film of water would reflect light, leach nutrients and encourage colonization by tiny epiphytes.

Epiphyte

Liana

Buttress root

Giant fern

Flowering
emergent tree

Emergent tree

Deciduous
canopy tree

Canopy

Understorey
of smaller
or younger
trees

Herb layer

Light gap

Smooth unbranched trunk

Climbing plant

Light-trapping mosaic – The rain forest canopy is composed of closely spaced trees with flat, spreading crowns. Individual crowns may be very large, supported on a simple pattern of branches which resemble the spokes of an umbrella. The crowns of neighbouring trees fit together like pieces of a jigsaw puzzle, but remain about a metre (3.3 ft) apart: a phenomenon known as "crown shyness". This spacing is not fully understood; it could prevent damage in high winds or the spread of plant-eating insect larvae. Looking closely at the mosaic pattern produced by such "shyness", it is possible to see mosaics within mosaics. The arrangement of leaves on a stem, and on the plant as a whole, is organized so that there is the minimum amount of overlap; each leaf is positioned so that it can absorb all the light it needs to photosynthesize efficiently.

Tree life cycles

Trees form the superstructure of the rain forest; it is their crowns that make aerial walkways for animals and roof gardens for epiphytes; their foliage that plunges the understorey into the cool, humid shade which nurtures seedlings and other delicate plants; their roots that collect scarce nutrients; and their mighty trunks that support the great weight of the canopy. The diversity of tropical tree species is immense. There are commonly 50–200 different species per hectare (2.5 acres), compared with ten species in a similar area of temperate woodland.

In spite of this diversity, many tropical trees look superficially alike and for this reason are difficult even for botanists to identify. It is therefore possible to generalize about their characteristics and describe a "typical" tree. The great majority of rain forest trees are about 30 metres (100 feet) tall, although some reach 60 metres (200 feet) or more. They have slender, unbranched trunks, smooth bark and a simple branching system (see pages 60–61). Few exceed one metre (3.3 feet) in girth, although giants 17 metres (56 feet) across have been found. The wood can be extremely hard and resistant to termites and other wood-boring insects. A few species, such as ironwood (*Eusideroxylon zwageri*), are so hard that even driving a nail into the trunk is difficult. The root system is very shallow and the tree is supported largely by buttresses or prop roots, which can extend five metres (16 feet) up the trunk. Many tropical trees are evergreen, shedding a few leaves throughout the year. The leaves are generally oval, undivided, thick and leathery.

An evergreen rain forest tree grows all year round and therefore has no annual growth rings. It may live anything from 150 to 1,400 years and usually takes 30–60 years to reach maturity and begin flowering. Even when mature, many do not flower and fruit annually, but only once in three to ten years. There may be an element of time-sharing between species using the same pollinators and dispersers. In addition, overproduction can be self-defeating, leading to population explosions of pests.

The obvious exceptions to this "typical" tree are those which are not evergreen. Semi-evergreen and seasonal forests (see pages 16–19) are common at the outer limits of the equatorial rain forest belt, in rain shadow areas and regions adjoining savanna. During the dry season, the deciduous trees not only shed their leaves but most of them also flower, so that the seeds are ripe and ready to germinate when the rains come.

Getting a start in life

Tropical trees, whether evergreen or deciduous, have similar requirements and tend to grow cheek by jowl. This density, together with their longevity, means that finding a growing place is one of the main problems that most trees have to contend with. They are not specialists, each species adapted to a different niche, but competitors over time and space. Although many are able to play the waiting game until they can take advantage of a gap caused by the fall of taller trees, others can only grow where there is plenty of light. Some species have devised ways of making space by usurping another's place or, strange as it may seem, by surrendering their place so that others of the same species may live. Strangler figs (*Ficus* spp.) are the usurpers *par excellence*, while at the other extreme are those species which flower only once, often *en masse*, and then die when their fruits have ripened. The talipot palm (*Corypha elata*) spends decades as a juvenile before producing its swan song: an inflorescence two metres (6.5 feet) long, bearing 60 million individual flowers.

After competition for space, the next most pressing problem for rain forest trees is that of eking out a living on nutrient-poor tropical soils. A number of tactics, such as rooting above ground or living in association with root fungi, are shared by many different trees (see pages 56–59). Also, it is no coincidence that a lot of trees belong to the pea family, Leguminosae, many of which have root nodules that contain bacteria capable of converting atmospheric nitrogen into a form the tree can use.

Buttresses (*left*) – Growing mostly on very shallow soils, many trees have little in the way of underground root systems. Instead, they are supported by roots which are largely above ground. In some species, these roots take the form of buttresses: thin strong flanges which may extend 5 m (16.5 ft) up the trunk. If these supports are cut away, the tree can easily be pushed over. In addition to providing support, the buttresses may assist in feeding. Instead of a tap root, the buttresses span a wide area, sending down fine feeding roots into the soil.

Mangrove prop roots (*above*) – Prop or stilt roots serve a purpose similar to buttresses (*left*), providing support in situations where underground roots are inappropriate. This type of root system is especially common in flooded and mangrove forests. It consists of stout aerial roots which arch out from the trunk and grow down towards the substrate, dividing on contact into a number of finger-like rootlets. The resulting scaffold of roots traps sediment, stabilizing the mangrove plant against the pull of the flood, or the flow and ebb of the tide.

Young leaves (*above*) – Tropical trees tend to produce new leaves in "flushes" rather than a few at a time. When this happens the tree looks from a distance as if it is flowering. The limp new leaves are borne in drooping clusters and are often completely different in colour from the older leaves – red, bright pink, bronze or white. The chemicals that cause these colours may protect the tender new leaves against strong light and deter herbivores until the leaves expand and darken into the characteristically tough, leathery mature foliage. In addition, producing a large number of new leaves all at once "floods the market", ensuring that most of the leaves reach maturity before they are damaged by herbivores.

Seedling strategies (*above*) – Some long-lived canopy trees produce large seeds. The rich food store in the seed enables the solitary seedling to shoot up – in some cases to a metre (3.3 ft) tall before putting out leaves. Others (for example, the dipterocarp trees of Southeast Asia) have winged seeds that fall to the ground *en masse*. This mass invasion swamps both predators and available space, ensuring that a large number of relatively small seedlings become established. Yet another strategy is used by short-lived trees such as *Cecropia* spp. (see page 65).

Strangler fig: usurping another's place
There are many different species of strangler fig (*Ficus* spp.) in the rain forest. Each species has its own specific wasp pollinator and fruits prolifically (100,000 fruits have been found on a single tree). The small seeds are dispersed by the many birds and monkeys that depend on the fruits for food. Now and again one of these countless seeds lodges in the branch of a tree and germinates. The seedling first sends out a long aerial root. When contact with the ground is made, the young fig starts to grow, putting out more roots from its perch to the ground, and developing stems and leaves. Eventually the host tree is smothered by the fig's foliage, its trunk is encased in its roots, and it dies. In this way, the fig avoids the competition on the ground, usurping the place of a tree which already stands tall.

Light gaps

From time to time, mature trees fall down, leaving a gap in the canopy which gives younger plants a chance to develop. They fall for a number of reasons: some are felled by people who live in the forest; others fall after being struck by lightning or high winds; but most fall simply because of old age, often hastened by termite damage or epiphyte burden. The gap torn through the forest is usually quite large, for a single tree, linked to its neighbours by immensely strong, elastic lianas, inevitably brings others down with it. At any one time the rain forest is a mosaic of ancient forest and light gaps that are in various stages of regrowth. Disturbance is the key to vitality and diversity in a habitat such as this that is otherwise extremely stable and dominated by long-lived species.

The opening of a light gap shakes the forest's equilibrium to its foundations. The dense ceiling of vegetation that normally shelters the interior of the forest, keeping it cool, humid, breathlessly still and in deep shade, is suddenly removed, allowing an influx of bright light and fresh air. In the gap the temperature is higher and the humidity lower than the surrounding forest, and as the damaged vegetation decays it releases nutrients. This gives a tremendous boost to the seedlings and saplings of primary forest species which may have been in a state of suspended animation for many years, waiting for such an event to continue their life cycles.

Animals soon move in, and before long there is a new influx of seeds from the surrounding forest. It is a highly competitive, dynamic phase. The young "jungle" grows quickly and in less than ten years has the same leaf density as that of the mature forest, although it has not yet reached the same height. During the first 15 years there is a rapid increase in species diversity; within 40–50 years, the gap will almost have closed, but the species present may be quite different from their predecessors. Diversity continues to build up, albeit slowly, even after a century has elapsed, until the gap finally becomes indistinguishable from its surroundings.

Many primary forest trees have adapted to the unpredictable way in which light gaps appear by having large seeds and shade-tolerant seedlings. Fuelled by the food reserves contained in the seed, the seedling establishes itself on the forest floor, but then undergoes painfully slow growth until a gap occurs and it receives sufficient light and warmth to burst into life and grow towards the light. Without this stimulus it can never reach its full size or reproduce, for the mature tree is dependent upon the bright light and life of the canopy. This type of tree predominates in the Amazon rain forest.

A catastrophic disturbance in the rain forest, such as a hurricane, volcanic eruption, earthquake or landslide, creates a swathe of destruction that is much larger than the gap caused by the fall of a single tree. The regeneration of large gaps is quite different from that of small gaps, involving rapid colonization by opportunistic, short-lived trees, such as the New World *Cecropia* species (members of the nettle family, Urticaceae) or the Old World *Macaranga* species, which in turn provide the necessary shade and humidity for canopy species to become established.

Cecropia trees do not reach a great height or age, seldom exceeding 18 metres (60 feet) and living only 30–80 years. They do however perform a vital role in the regeneration of large gaps, providing quick cover for the vulnerable root mat, encouraging animals to visit and take up residence in the disturbed area, and creating the necessary shade in which an understorey of primary forest "small gap" saplings can develop. As the pioneering *Cecropia* trees die out, small gaps are created which allow the longer-lived species to develop.

Catastrophic disturbance by natural means, whether it creates a small or a large gap, is quite different in its effects on the forest from selective or clear felling (see pages 182–183), or from large-scale burning. Unlike fire and heavy machinery, natural disturbances do not destroy the root mat, which in many cases is all that stands between the forest and desert.

Montane light gap (*far left*) – In this montane rain forest in Venezuela, especially where the canopy is fairly open, the appearance of a light gap does not affect the environment within the forest as much as it might in a closed-canopy lowland forest. Saplings and shrubs grow to fill the space left by the fallen tree which, because of the lower temperatures found at altitude, decomposes comparatively slowly.

Reaching for the light (*left*) – In the tangle of lianas that connect many of the trees in this rain forest in Southeast Asia, a falling tree usually brings down others with it. At once the saplings present in the gap respond to the increased light levels and put on a spurt of growth in an attempt to reach the canopy.

Filling the large gaps

Cecropia trees fruit often and heavily. A single tree produces about 900,000 small seeds which are widely dispersed by fruit bats at night and birds during the day. The seeds remain viable for two years, and are present in soil throughout the forest in surprising numbers – an average of 73 seeds per square metre (11 square feet) has been recorded. Germination is rapid when a large gap provides suitably high temperatures and bright light. The young trees put all their energy into growth, investing little in defensive measures such as hard wood and toxic leaves. The trunks are hollow and the wood so light that it is often used for floats by local people. *Cecropia* foliage is palatable, providing a staple diet for many different animals, including sloths and monkeys.

Competition in large light gaps is fierce, particularly from juvenile climbers which constantly threaten to overwhelm young trees. *Cecropia* beats the opposition by employing gardeners, in the form of non-stinging *Azteca* ants. The ants are housed in the hollow branches and feed on drops of nectar exuded from special nectaries on the leaves, and on food capsules, known as Müllerian bodies, which are situated at the base of each leaf stalk. The capsules contain protein and glycogen, an animal starch which *Cecropia* alone among plants is known to produce. In return, the ants patrol the tree vigilantly, severing the tendrils and twining stems of invading climbers, and throwing overboard any germinating epiphytes. Thus unhampered, the young trees grow 2.5 metres (eight feet) a year.

Azteca ants (*above right*). *Cecropia* tree (*right*).

Epiphytes: plants on plants

In almost any temperate forest some mosses and lichens are ordinarily visible on the bark of the trees. These epiphytes – plants that live on other plants – often form a thin layer over the surface of their host. What is different about the rain forest is the number and diversity of epiphyte species. In addition to the usual mosses, lichens and ferns, there are orchids, cacti, bromeliads (members of the pineapple family, Bromeliaceae), aroids (plants of the arum family, Araceae) and representatives from numerous other plant families. So numerous are the epiphytes on many trees that their leaf area may exceed that of the tree on which they are growing. Epiphytes are not parasitic, in the sense that their roots do not penetrate the host plant's tissues, but they do compete for light and nutrients, and may contribute to the demise of the host by their burden.

A quarter of all plant species in lowland rain forest are epiphytes. Montane forest may have an even greater proportion, due to the exceptionally high atmospheric humidity which encourages their proliferation. Indeed, almost all montane forest plants, other than the trees, seem to be epiphytic, for every surface has a thick covering of moss in which other plants may then take root. Tropical America alone has an estimated 15,500 epiphytic species. African rain forests have fewer epiphytes than other rain forests, perhaps because many of them became extinct during the successive dry periods that were a feature of the last Ice Age, whose impact was more severe there than elsewhere. Even so, 47 different species of orchid have been recorded on a single West African tree.

Like other rain forest plants, epiphytic species tend to occur in strata. Canopy epiphytes are exposed to sun, wind and occasional dry periods. As a result, they show many of the same adaptations as desert species: foliage that is either thick and leathery or very narrow to prevent dehydration and scorching, and extremely efficient ways of absorbing and storing water. In contrast, epiphytic species in the understorey have to contend with low light and permanently damp conditions. Consequently, they often have thinner leaves, "drip tips" (see page 60) at the ends of the leaves to get rid of excess water, and corrugated or velvety textures which increase the surface area and thus the light-gathering capacity of the leaf.

The epiphytic way of life is precarious. The host tree may shed its bark periodically to rid itself of epiphytes, or it may lose limbs under the weight of wet vegetation (which may total several tonnes per tree). Eventually it will die and decay. Such events spell doom for most of the epiphytes in the higher reaches, for they are unlikely to survive the fall or the damp, shady conditions on the forest floor. But these hazards must be weighed against the advantage, in a habitat dominated by tall plants, of gaining access to the light without investing in lengthy stems.

Survival techniques

The success and abundance of epiphytes in the tropics depends on a number of factors. High atmospheric humidity, more than high annual rainfall, is essential to prevent drying out. The plants must have a secure means of attachment. There are often two different root systems, one that clasps firmly to the host, and another that penetrates debris or grows freely in the air to collect moisture. Epiphytes need ingenious means of making the most of irregular supplies of water and nutrients. Many have swollen stems or leaves arranged like funnels to collect water. They must also be prolific and adaptable in their reproductive strategies. The right conditions for flowering may be few and far between, so many species have adapted to this by reproducing asexually in the meantime.

Other survival techniques include living in association with root fungi (mycorrhizas), which provide essential nutrients and especially at the seedling stage, or with ants, which pack grains of soil and debris around the roots to make their nests (*far right, bottom*). Ant-garden epiphytes include aroids, bromeliads, gesneriads, epiphyllums and peperomias.

Fine seed (*right*) – The seeds of orchids such as *Stanhopea tigrina* are as fine as powder. Although the seed is widely dispersed by the gentlest of breezes, suitable growing places are scarce and the minute seeds have no food reserves. Successful germination and growth depend upon the presence of root fungi which provide vital nutrients. Orchids are slow to grow and take years to reach flowering size. Even then, conditions for flowering – such as a pronounced dry period – may not occur annually. In the meantime, the plant invests a lot of energy in producing new growths, each of which is capable of independent existence, should it be severed from the parent plant. Such new growths improve the chances of pollination in that they produce a more noticeable display of flowers to attract the orchid's insect pollinators.

Litter basket epiphyte – A number of epiphytes, including the giant aroid *Anthurium salviniae*, grow as a rosette of leaves which funnels rain, dew and debris into the centre of the plant. Although superficially similar to the vases of epiphytic bromeliads, the rosette contains an arboreal compost heap rather than a reservoir of water. Two kinds of roots are produced: adhesive, clasping roots which attach the plant firmly to the host tree; and short, upward-growing roots which penetrate the damp, decomposing litter, anchoring and aerating the debris, and tapping the supply of moisture and nutrients.

Gardens in the air (*above*) – Epiphytes do not live entirely at their hosts' expense. Their foliage and root sytems trap debris and moisture, building up an "epiphyte mat" along the host tree's branches and providing many hiding and breeding sites for small animals, such as insects and frogs, some of which live their entire lives in these aerial gardens. The mat is also a source of moisture and nutrients. Some trees actually send out aerial roots to tap this resource. The only thing that significantly reduces the numbers of epiphytes on a tree is tree travel (see page 88–89). Most animals use well-established routes through the canopy, and the upper surface of a branch which bears regular traffic is clad in nothing more than a padding of moss – although the sides may still bristle with larger epiphytes.

Ant-friendly epiphyte (*right*) – The Indonesian epiphyte *Myrmecodia echinata* clings to trees with its roots and develops swollen, tuber-like lower parts containing chambers that are inhabited by ants. The ants deposit organic matter in the chambers which they inoculate with fungal spores. As this matter decomposes, nutrients are released which feed both ants and plant. In addition, the ants disperse the plant's seeds and guard the plant against invasion by other epiphytes and animals. In exchange, the plant supplies the ants with nectar and a place to live.

Climbing plants

Impenetrable undergrowth and loops of lianas (woody climbing plants) are part of everyone's idea of a jungle. It is the lianas and non-woody climbers, untidily slung between the trees forming confused patterns, that confer the air of mystery often associated with rain forests. Although climbing plants occur in most plant families, the majority are found in the tropics. They have evolved very successful ways of competing for light in a habitat dominated by tall trees – not, like epiphytes (see pages 66–67), by perching on their competitors, but by winding and clasping their way up them.

Reaching for the light

About eight percent of the plant species in lowland rain forest are climbers, the highest percentage occurring in secondary forest. They show several different approaches to the problem of getting to the light. The simplest is to have a barbed stem and to use brute force, thrusting upward and gaining purchase by thorns which hook on to the surrounding vegetation. Rattans (climbing palms), which can reach 200 metres (650 feet) in length, use this technique. They also have barbed, whip-like extensions to their new leaves which do not unfurl until the "whip" has lashed itself to the next support. Pitcher plants (*far right, top*) also climb by having "whips" on the ends of their new leaves, which swell into pitchers as the leaf develops.

An alternative approach is used by the climbing aroid group (genera of the arum family Araceae, which include *Philodendron*, *Monstera*, *Rhaphidophora* and *Scindapsus*). They produce two kinds of roots: short ones put out at right-angles to the stem, which develop adhesive hairs when they make contact with the climbing surface, and long feeding roots which may dangle in mid-air initially, but branch profusely as soon as they encounter the soil. The clasping roots often attach the plant so securely that it takes considerable force to tear them away. The aerial roots are extremely flexible and strong. When anchored, they act like guyropes, tethering the host tree and playing an important role in the structure of the forest (see pages 64–65). Local people use the roots of many different species for cordage, from fine weaving to heavy-duty ropes. Some climbers of this kind may become almost epiphytic. Although they germinate and begin growth on the forest floor, they may eventually flourish entirely in the tree tops, losing connection with the ground as the original roots die.

Another approach is to put out tendrils from leaves or stems. This is common among lianas such as *Leea* spp., *Cissus* spp., *Tetrastigma* spp. (the host plant of the parasitic *Rafflesia* spp.), *Bignonia* spp. and *Passiflora* spp. An outstretched tendril moves away from the light and makes sweeping movements as it searches for a support. Brushing against another plant stimulates it to curl. The response is rapid: the tendril of one tropical American gourd curls within 20 seconds of contact and completes its first coil in four minutes, the tissues thickening as it does so to strengthen its grip. As most tendril climbers are attached only at their extremities, the heavy water-conducting stems tend to sag and loop through the vegetation. Some species have stems filled with water, which can be tapped by forest peoples.

A number of tropical climbers have different juvenile and adult forms. Juvenile foliage may differ from the adult in size, shape or colouring. The advantages of such differences are difficult to determine, although one possibility is that an insect searching for a particular plant on which to lay its eggs may not recognize the juveniles and so the young plant is spared until mature, when it is less vulnerable. In several species of *Monstera*, *Cercestis*, *Ficus*, *Hoya* and *Marcgravia*, the juveniles are known as shingle plants because their leaves overlap like tiles on a roof. This habit reduces water loss to a minimum. The leaves are small, clamped to the host tree's bark and have pores only on the undersurface, so any moisture that is lost remains trapped under the leaf. As soon as shingle plants reach the light, the foliage changes dramatically, developing long stalks and large blades: flowering and fruiting can then begin.

Pitcher plants – Of a final height approaching 9 m (30 ft), and with pitchers as much as 30 cm (12 in) in length (which are usually crammed full of digested insects), *Nepenthes rafflesiana* is one of the most formidable of pitcher plants. Although insects are the mainstay of its diet, small mammals and reptiles also fall prey as they attempt to cash in on the plant's bounty.

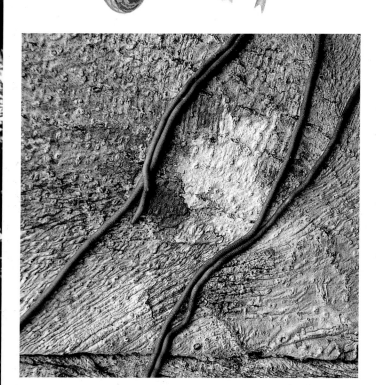

Dangling lianas (*left*) – The stems and feeding roots of lianas dangle from the trees on which they are attached. Once the roots have grown down to the ground, they anchor themselves in the soil. Tethered and chained together by looping stems and strong flexible roots, rain forest trees both gain and lose from this close embrace. Although more stable in high winds, they are also more likely to fall in an inextricable tangle if a neighbouring tree is uprooted or felled.

Clasping its host (*above*) – In addition to aerial feeding roots, climbing plants such as *Philodendron* spp. put out clasping roots which attach the plant firmly to the host tree. These sinuous roots mould themselves to every contour of the bark. Sometimes they branch and fuse wherever they cross, forming a lattice that encases the tree trunk.

Shingle plant (*above*) – Rain forest climbers germinate on the forest floor. They then not only have to compete with the surrounding plants, but must also find a suitable host to support their growth. The most ingenious solution to this problem is found by *Monstera dubia*. It produces an exceptionally large seed which puts out a leafless, rootless, cord-like stolon. As it grows, the stolon is attracted to dark shapes, such as a host tree. The "search" for a host can extend up to 2 metres (6.5 feet) before its food reserves are exhausted. As soon as the stolon reaches a host it turns into a shingle plant and climbs upwards. During this stage *Monstera dubia* is adapted to conserve moisture, for it is not able to gather much water and may be sheltered from rain in its position against the tree trunk. When light levels are high enough, it produces large, long-stalked, leathery, mature leaves.

Flowering and fruiting

Two-thirds of the world's flowering plants are found only in the tropics. One reason for this tropical diversity is the large number of relationships they have forged with animals (see pages 76–79). In the rain forest, the web of relationships between animals and flowering plants is so complex that the loss of one species can lead to the extinction of many others. Another reason for this diversity lies in the fact that the climatic conditions in a rain forest vary little throughout the year, or even over centuries. This has allowed the evolution of a stable community of long-lived perennial plants, in which long periods elapse between generations, and the opportunities for new individuals to become established are few and far between. Consequently, competition between juveniles is intense; smaller trees, climbers and epiphytes all face the same uphill struggle.

There are three main strategies that a plant in this situation can adopt to gain an edge over its rivals. First, it can multiply its chances by reproducing sexually so that its offspring are as genetically diverse as possible. The majority of the flowering plants in the rain forest cannot fertilize themselves: to reproduce they must be cross-pollinated. (At the opposite extreme are habitats in which conditions are unstable. These have a high proportion of self-fertile annuals.) Cross-pollination depends largely on insect or animal pollinators because there is little wind in the forest to disperse pollen. There are a couple of exceptions to this rule. A number of canopy epiphytes, such as orchids, do use the wind to disperse their powder-fine seeds. The second exception is the massive dipterocarp trees that dominate some of the Southeast Asian rain forests, pushing their crowns high above the canopy. Unless whipped up into the air by storm winds, their winged seeds fall to the ground only a short distance from the parent tree – not surprisingly, dipterocarps spread through the rain forest extremely slowly.

The second strategy that a plant can adopt is to produce a mass of seeds to ensure that its genes have as many opportunities as possible for expression. Flowering and fruiting is therefore a major investment in terms of energy, and takes place only when conditions are right. Because many trees take 30 to 40 years to mature and seldom flower annually, when flowering and fruiting does take place it is often spectacular. A hectare (2.5 acres) of lowland rain forest produces as much fruit and seed, by weight, as 12 hectares (30 acres) of temperate oak woodland.

The third strategy is to ensure widespread dispersal, so that if the conditions are unfavourable in one location a few of the seeds have a chance of succeeding elsewhere. Most plants use highly mobile animals – such as birds, bats and monkeys – for this.

Tropical rain forest plants are not alone in having complex flowering and fruiting cycles, but they do have certain characteristics that are either rare or unknown in other habitats. A number of lower canopy trees produce flowers (and subsequently fruits) directly from their trunks and branches, a phenomenon known as cauliflory. Many cauliflorous species are pollinated by bats which may find it easier to reach flowers that are positioned clear of foliage. The cocoa tree (*Theobroma cacao*) is an exception: it is pollinated by midges. However, the large cauliflorous fruits do not fall when ripe, but are opened *in situ* by squirrels and monkeys, which can easily get to the fruits along the trunk and branches. The decaying pods provide breeding sites for the midges, which are then on hand for further pollination.

Attracting pollinators
Many tropical plants produce very large flowers. Those of the balsa tree (*Ochroma lagopus*) are 12 centimetres (five inches) long and eight centimetres (three inches) across at the mouth. Sizeable fleshy flowers are often necessary to withstand the attentions of

Orchids (*above*) – There are more than 9,000 species of epiphytic orchid in the tropics, most of which have restricted distributions. Many are pollinated by a single species of insect, or just one sex of a species, which makes them among the most specialized of flowering plants and particularly vulnerable to extinction. Orchid flowers are complex in structure and produce a wide range of scents, from sophisticated perfumes to carrion stenches.

"Hot" pollination

To ensure successful pollination, *Philodendron bipinnatifidum* has evolved one of the most sophisticated pollination strategies found in the rain forest. First, it attracts its insect pollinators by producing a powerful odour which can be detected at great distances; then it persuades the insects to stay long enough to complete pollination by offering them ideal conditions for mating. Flowering takes place over several months, but each day only one or two inflorescences open. The inflorescence opens at dusk and lasts for 24 hours only. Separate zones of male and female flowers are arranged at the base of a poker-like spadix, which is surrounded by a cowl-shaped spathe. There is also a zone of sterile male flowers which heat up the spadix to as much as 17°C (31°F) above the ambient temperature, volatizing the pungent odours which summon scarab beetles of the species *Erioscelis emarginata*. The fuel that powers this increase in temperature is in the form of lipids – fatty or waxy substances that are common in animals but unknown elsewhere in the plant world.

To produce the required amount of heat, *P. bipinnatifidum* consumes nearly as much oxygen as a flying hummingbird, which even by animal standards is extremely high.

Once inside the spathe chamber, the beetles are offered food (in the form of the sterile flowers) and a warm, odorous place in which to mate. Thus occupied, they remain there until pollination is complete. By dusk the next evening, the spadix cools down, the spathe closes around the pollinated flowers, and the pollen-covered beetles fly off to a newly opened, warm, scented inflorescence for another orgy.

Living fossil (*left*) – Before the flowering plants became the dominant plant type in the rain forest, conifers, cycads and tree ferns were common. Another plant that was present then, and survives to this day, is the *Gnetum* vine. It produces clusters of "flowers", resembling upside-down catkins. Each flower is protected by a petal-like frill of bracts. After pollination, the bracts of the female flowers grow fleshy to attract seed dispersers.

large pollinators, such as bats (as in the case of the balsa tree) or birds, although the world's broadest flower is in fact pollinated by flies – *Rafflesia arnoldii*, a parasitic plant endemic to Sumatra, has thick leathery flowers a metre (3.3 feet) or more in diameter.

Brilliant colours are also common in the rain forest, especially in flowers pollinated by birds, which for the most part have excellent colour vision.

Other pollinators may be drawn to flowers by strong scents that can be detected at some distance in the still, humid air. A great many rain forest species are pollinated by moths, for the nights are always warm enough for insects to be active. Their nocturnal flowers emit intense perfumes, reminiscent of scented soap, and are commonly white or pale in coloration, with dissected outlines that show up well in the dark. The best-known example is the Madagascan orchid *Angraecum sesquipedale* (see page 147).

Some plants have equally powerful, unpleasant scents which attract carrion beetles or flies. The flowers are often dark purplish-brown in colour and mottled, resembling decomposing flesh. *Aristolochia* vines produce elaborate fly traps which smell like rotting fish. The flies are taken in by the flower in more ways than one; they not only lay eggs on the dummy corpse but also fall into the trap where they are detained until the female flowers are pollinated and the male flowers have dusted them with pollen. The orchid *Dracula bella* also fools its pollinators. The lip of the flower mimics the underside of a fungus and attracts fungus gnats which, in the process of laying eggs among the fake gills, carry out pollination.

Although some plants achieve pollination by deception, the majority reward their pollinators, some with nectar or pollen that may be varied in consistency or chemistry to suit specific pollinators. Bat-pollinated flowers usually have sticky, rather than runny, nectar that is easier to lick, and high-protein pollen, rich in substances that are essential for the growth of wing and tail membranes. Many rain forest species produce other "designer" substances. Waterproofing substances such as waxes and resins are especially important in the wet tropics. *Clusia uvitana*, a Central American shrub, coats its flowers in wax which is collected by *Trigona* bees for nest-building. *Amorphophallus variabilis*, which grows on the ground in Southeast Asian forests, produces special food bodies filled with oil and starch in a bowl-shaped inflorescence. The pollinating beetles are engrossed in feeding for several days while the female and male flowers complete their cycles. Keeping the large clumsy insects immobilized prevents the flower's vital parts from being damaged by the beetles' sharp claws and spines.

Fragile interdependence

In many cases, plants and their pollinators and seed dispersers have evolved so closely together that the loss of one species spells doom for all the others in the chain. For example, more than 20 species of bird depend on the fruit of the Costa Rican tree *Casearia corymbosa* for several weeks each year when virtually no other suitable food is available. However, the tree's seed is dispersed only by the *Tityra*, a small bird whose conspicuous silvery plumage and black facial mask makes it highly conspicuous and easy prey for hawks. Its feeding strategy is therefore quite different from that of the other birds. Instead of staying in and around the tree, it makes quick dashes and flies off to consume the fruit elsewhere, thus depositing the seeds some way off. *C. corymbosa* and *Tityra* depend on each other; but the decline or extinction of either would also have a knock-on effect, reducing or eliminating the other birds that feed on the fruit, and in turn other plant species that depend upon the birds as dispersers (or pollinators). This feature of the rain forest ecosystem makes it especially vulnerable to disturbance and a particularly difficult environment in which to assess the full consequences of the loss of a single species.

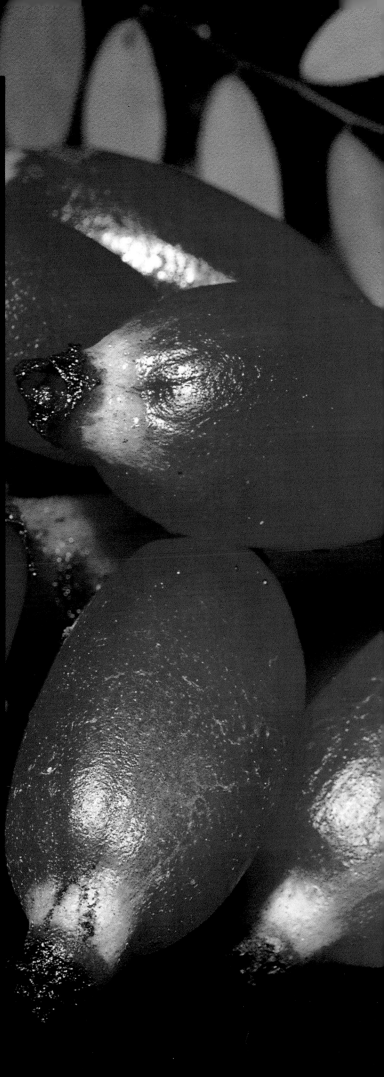

(*Main pic*) Wild ginger fruits become brightly coloured when ripe to attract animals to disperse the seeds.

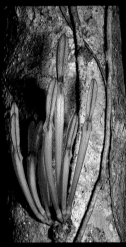

Colour for birds – The Sarawak mistletoe (*Macrosolen* sp.) is one of more than 1,000 species of parasitic mistletoes in the tropics. Many have a close relationship with flowerpeckers, small birds of the family Dicaeidae. This mistletoe species has red flowers to attract the birds and petals which are sensitive to touch, springing open to allow access to the nectar.

Odour for bats – The flowers of the calabash tree (*Crescentia cujete*) open at night, releasing an odour like sweaty cheese which attracts bats. Bat-pollinated flowers are usually pale or dingy in colour, for most bats do not use sight to find their food. The flowers of both the mistletoe (*above left*) and the calabash grow directly from the trunk which gives easier access for bats or birds.

Water everywhere

It rains in the tropical rain forest as it rains nowhere else on earth. Torrential downpours that wash nutrients from leaves, that soak bark, wood and soil to saturation point, and that drench the forest inhabitants and make the air heavy with water vapour. The usual distinctions between water and dry land break down in this strange world, producing fish that live on the forest floor and crabs that live in trees.

The high humidity of the air inside a rain forest means that animals with thin, moist skins are rarely in danger of drying out. Invertebrates that are normally found only in water may be seen slithering over the vegetation of the lower storeys. They include colourful flatworms, and the rather less appealing leeches, which lurk among the leaves and attach themselves to passing animals. In the leaf debris around epiphytic plants, earthworms live out their lives hundreds of feet above their normal habitat.

Humid air also favours frogs, which flourish in the rain forest in exuberant variety. Some have become so completely terrestrial that they have no need for pools of water at all, not even for breeding. Their eggs, laid on leaves or on the forest floor, do not dry out as they would in any other habitat. The tadpole never emerges from the egg, but develops into a tiny frog inside its coat of jelly.

Pools in the sky

Other rain forest frogs use the pools of water caught in epiphytic plants for rearing their tadpoles. These stores of water, high up in the canopy, are an important feature of the rain forests. In the Americas, bromeliads are the major water-storing plants, catching rainfall within a circle of stout, prickly leaves. Living on open branches high up in the canopy, where the humidity is much lower, they sometimes suffer drought conditions between rainstorms. Their miniature pools help to even out the water supply, and may also supply them with extra nutrients (see page 68). A variety of animals live in these miniature pools-in-the-sky, including the larvae of mosquitos and damselflies. Arrow-poison frogs bring their tadpoles to the bromeliad once they have hatched, but cannot afford to lay their eggs there for fear of predatory insect larvae. Although a microcosm of life, with its own complex food webs, the bromeliad pool does not contain enough nourishment for the tadpole. So the mother frog returns each day to lay an infertile egg which the tadpole consumes.

On the other side of the world, in the montane forests of New Guinea, an unrelated tree frog has been found guarding a mass of eggs in the water-storing tubers of a *Hydnophytum* epiphyte. In another intriguing parallel, a crab that breeds in pitcher plants has been discovered in Malaysia, and another species that rears its brood in a bromeliad was recently found in Jamaica. Like the arrow-poison frog, this crab tends its young and brings them food.

Swimming through the canopy

The hazy dividing line between land and water becomes yet more blurred in the flooded forests of the Amazon Basin, where the water level rises by up to 12 metres (40 feet) annually, high enough to submerge many trees. Surprisingly, these do not lose their leaves. Some of the trees in the flooded forest – the ones that are not inundated – actually rely on fish to disperse their seeds, and produce floating fruits that are eaten by fish in huge numbers (see page 113). This phenomenon is not confined to the Amazon Basin. In Malaysia, too, there are fruit-eating fish, which act as seed dispersers, expecially in the rivers that drain the peat-swamp forests.

When the waters fall again from the flooded forest, they leave behind freshwater sponges stuck firmly to the trunks of the trees. Protected by a tough outer layer, these sponges stay there in a state of suspended animation, until the floodwaters rise again the following year. Also left behind is a tiny blood-red catfish, about a centimetre (half an inch) long, which lives among the leaf litter. Blind and scale-less, it finds its invertebrate prey with the help of sensitive whiskers.

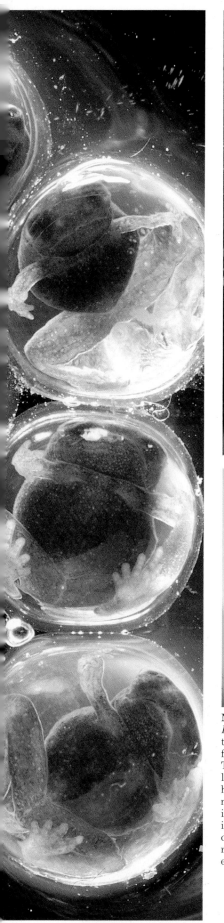

No need for water (*left*) – *Eleutherodactylus* frogs complete their development from egg to fully-formed frog on dry land. The eggs are usually laid on leaves or moss. The tadpole never has the chance to swim before it metamorphoses into a froglet; instead it is confined to wriggling inside its transparent egg capsule. When the tiny froglet is ready to emerge, it uses a small egg-tooth to cut its way out.

Attracting females (*top*) – The colour of these male golden toads (*Bufo periglenes*) is believed to help the females find them in the gloom of the forest. In contrast, the female (bottom left of picture) is a dull colour.

Epiphyte frog (*above left*) – A red-and-blue arrow-poison frog (*Dendrobates pumilio*) completes its metamorphosis from tadpole to frog in an epiphyte pool.

Hitching a lift (*above*) – Once hatched, the tadpoles of the poison dart frog *Phyllobates lugubris* wriggle on to the back of the male frog and are carried to water, normally in a forest stream or bromeliad pool, to complete their development. The mucus secretion with which they are stuck to the back of the male dissolves away in water, leaving the tadpole to swim free.

Fruit and seed eaters

In the rain forest, an opportunity for a canopy tree to grow to its full height comes only occasionally. Taking advantage of these rare opportunities means producing as much seed as possible, and dispersing it well, so that some seeds or seedlings will be in the right place at the right time. The vast majority of rain forest plants rely on animals to disperse their seeds, wind being of little use in the stillness of the forest. Fruits are a bribe that plants offer animals in exchange for the advantages of mobility. In the simplest transaction, the animal eats the fruit then deposits the seed unharmed, some way off, allowing it to germinate at a distance from the parent tree. But things are rarely quite this simple because the plant's seed is also nutritious, often more so than the fruit. The plant must not only cope with animals which feed on the fruit without dispersing it, but also with the "seed thieves" which want to eat the seed, and with animals that disperse it but inadvertently destroy its ability to germinate, usually through the action of their digestive system on the seed.

One solution to these problems is to produce hundreds of gritty little seeds packed into a juicy, easily digested fruit that attracts all types of frugivores. The fruit needs almost no chewing or crushing, so most of the seeds are swallowed intact, and, with luck, survive their passage through the digestive system. This is the strategy that figs adopt, and it serves them well, but for the seeds to survive they must germinate where there is a fair amount of light. Unless they can quickly become self-sufficient they will perish because the food reserves in their tiny seeds are strictly limited. Most fig seeds are not so fortunate, but by producing millions of seeds each year the figs win through. This means the bearing of prodigious amounts of fruit, and figs are the mainstay of fruit-eating animals in many rain forests.

For most rain forest trees, the fig's strategy will not work, and a large nutritious seed is essential. Such trees tend to specialize in attracting a particular type of fruit eater – one that will disperse the seed rather than destroy it. By focusing its efforts on a particular fruit eater, the plant can defend the seed against that animal's digestive system. Some form of hard seed coat may protect the seed when inside the animal, or a cocktail of toxic chemicals may be employed.

Directing the fruit to the chosen disperser involves a carrot-and-stick approach, attracting the favoured animals and deterring the others with chemical or mechanical barriers. Such devices have evolved during millions of years of gradual coevolution with animals in the forest, but there are too many potential enemies for any plant to perfect its defences. Even the most powerful poisons are not invincible, and seed eaters such as the black colobus monkey (*Colobus satanas*) of West Africa can detoxify a wide range of chemicals that are poisonous to other mammals. Fruit bats thrive on doses of strychnine, arsenic and cyanide that would be fatal to a human. It is a battle that the plants can never hope to win outright, and most species lose a large proportion of seeds, but generate enough for a lucky few to survive.

The most useful dispersers are those that fly – the birds and bats – because they are more likely to carry the seed some distance away. Bird-oriented fruit tends to be brightly coloured, often red, and to split open when ripe so that birds can peck out the sweet flesh with its attached seeds. Bats are attracted with scents rather than colour because they fly by night. Bananas are typical bat fruit, with their dull colours, thick skins, and pungent smell when ripe. Musty smells are also attractive to bats, and many of their fruits are distinctly unappealing to the human palate.

Some of the largest canopy trees put out a seed that is too big for a fruit-eating bird or bat to disperse. A common strategy is to drop these giant seeds to the ground for large animals to pick up. In the African rain forests, elephants disperse the seeds of some trees, gathering around trees and gorging themselves on the fallen fruit. Some Amazonian trees, such as the cannonball tree

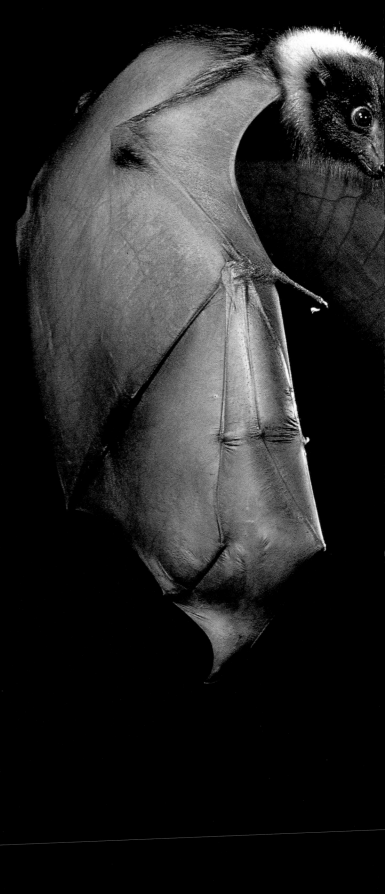

Colourful fruit eater – The resplendent quetzal (*Pharomachus mocinno*) eats many fruits including those of the wild avocado. The hard-shelled stone of the avocado fruit passes through the bird's gut unharmed, to germinate in its droppings.

Doubly useful – The bill of the keel-billed toucan (*Ramphastos sulfuratus*) is useful for reaching fruit, but also plays a part in displays. The toucans of the New World and the hornbills of the Old World are an example of convergent evolution.

Intent on its prey (*main pic*) – Like other fruit bats, the Indian fruit bat (*Pteropus giganteus*) uses mineral salts from seawater to supplement its diet of fruit. Fruit bats are found mainly on islands and in coastal areas.

(*Couroupita guianensis*), drop their fruit in the same way. But for some unknown reason the fruit sits in piles and rots. Perhaps it was once dispersed by a large animal that has since become extinct, possibly a giant ground sloth or the elephant-like mastodon. Where the original dispersers have vanished, some strange relationships between trees and dispersers have developed (see page 55) in their place.

The harvest of fruits in lowland rain forests is so rich that a hornbill can gather 24,000 fruits on which to feed his mate and chick during the breeding season. Specialist birds and bats, which eat almost nothing but fruit, flourish in the tropical forests as nowhere else in the world. They include the toucans of South and Central America, the palm nut vulture (*Gypohierax angolensis*), which feeds on oil palm and raffia fruits in the freshwater swamp forests of West Africa, and the spectacular pied imperial pigeon (*Ducula bicolor*) of Southeast Asia, also known as the nutmeg pigeon because it relishes the fruits of the nutmeg. These are swallowed whole, the large seed being voided in the birds' droppings. Some species can swallow fruits that are slightly larger than their own heads, thanks to highly elastic joints linking the upper and lower jaws. The tough outer coat and aril of the nutmeg are scraped away from the seed by the pigeons' specialized gizzard, which is lined with horny "teeth".

Most fruit eaters need no such specializations, for fruits are generally easy to digest. Short digestive tracts are the norm, and some birds extract the nourishment from fruits so rapidly that they void the remains in just five minutes. The giant fruit-eating bats of Southeast Asia, known as flying foxes, eat a largely liquid diet. They crush the fruit like a strainer, retaining the pith, peel and seeds, which the bat then spits out.

The ease with which fruit can be gathered and eaten allows plenty of time and energy for other things, such as courtship displays. Sexual selection favours such displays, particularly among birds, but in most habitats it is countered by the need to find food, which sets limits to the time that can be invested in display. Tropical fruit eaters such as the birds of paradise, bowerbirds, manakins and cotingas know no such restraints and have developed displays of unparalleled colour and passion.

Threats to survival

Seed dispersers are the key to the survival of the forest. Where rain forest has been cleared for pasture land in Amazonia, areas with some standing trees (even if these are dead) begin to regenerate soon after being abandoned. Those with no trees remain grassland, because without a place for birds and bats to perch, no seeds are brought in from the forest. In other areas, the forest remains but some of its trees may already be doomed to extinction because their dispersers are gone. Some isolated rain forest remnants in Southeast Asia have lost all their hornbills through hunting, and the trees now standing in them may well be the last generation.

Although the fruit crop of the rain forests is abundant, it is not evenly spread throughout the year. There is a season of glut and a season of relative scarcity. It is those plants that produce fruit in the scarce season that are crucial to the survival of the specialized fruit eaters. In a great many forests, the figs fulfil this role, because they fruit all year round. The strangler figs are especially important, but they are badly affected by logging, particularly in Southeast Asia. They have the misfortune to attach themselves to large trees, such as dipterocarps, which are highly attractive to the timber trade.

Even selective logging, which is generally favoured by conservationists, can destroy up to 75 percent of the fruit trees, primarily strangler figs. Without these, many fruit-loving animals must die out, but if mature forest is near by, and the logged areas are left to recover, both fruit trees and their animals will come back. In Malaysia, it has been shown that after ten years without human disturbance a logged rain forest can once again sustain a full complement of eight hornbill species.

Time to show off (*main pic, inset*) – Some of the most spectacular fruit eaters are the birds of paradise. The New Guinean rain forests are the centre of their distribution, but there are also 4 species in Australia. The courtship displays of these extraordinary birds are difficult to photograph, since they generally take place in dense, dark forest. However, some species remove leaves immediately above the branch on which they are about to start displaying, to provide a spotlight of sun. Birds of paradise were once relatively common, but hunting for hat-makers in Europe severely depleted their numbers. In the late 19th century, the feathers from about 50,000 birds were being exported every year. The trade was banned in the 1920s.

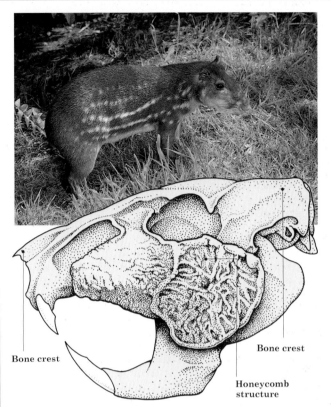

Bone crest

Bone crest

Honeycomb
structure

Cracking the nut

The seedlings of trees that are destined to reach the canopy of the rain forest germinate in the gloom of the forest floor, but must then play a waiting game, dependent for further growth on a break in the canopy above. To get to the stage where it has enough leaves to maintain itself in this twilight, the seedling needs a generous nutritional subsidy from the parent tree, in the form of a large, oil-rich seed. Trees like the Brazil nut (*Bertholletia excelsa*) equip their offspring with such a seed, but then face the problem of warding off hungry seed thieves. Their answer is to encase the seed in a rock-like seed-coat to keep out these marauders. But in the evolutionary arms race, almost every defence is overcome, and animals such as the peccaries and agoutis have evolved the means to penetrate even the hardest seeds. The peccary crushes hard seeds between its large flat molars. In the case of the agouti, the long incisor teeth can gnaw away at hard seed cases. Once the shell is pierced, the agouti can push its teeth inside and lever the remains apart. The teeth are powered by massive jaw muscles that are anchored to special crests of bone on the back of its skull and tip of its snout. If this formidable apparatus were made of solid bone, the agouti's head would hang heavily downwards, so parts of the skull are honeycombed to lighten the load.

Ironically, some trees have battled with the agouti for so many thousands or millions of years that they are now dependent on it to release their seeds from the heavily fortified cases. The Brazil nut is one such species and another is the leguminous tree, *Hymenaea*, whose sturdy pods firmly imprison the seeds. The agouti gnaws through the protective covering around a parcel of seeds, eats some of them and then buries the rest. Because it never rediscovers all its caches, a reasonable proportion of the seeds survive to germinate. The trees that depend on agoutis for dispersal tend to produce their seeds all at once, which encourages cacheing by the rodents.

Nectar and pollen eaters

Nectar is food at its simplest – sugar dissolved in water, providing a diet that needs no digestion whatever. Rain forest trees and vines provide nectar to tempt animals for pollination (see pages 70–75) and feed ants at "extra-floral nectaries" (see page 67). Pollen is a secondary food for many nectar eaters, providing a useful supplement of protein and nutrients. Plants whose pollinators eat the pollen itself, as well as taking nectar, produce extra pollen to compensate for such losses. Because pollination is so difficult in the rain forest, nectar and pollen are produced abundantly, and pollinators span the range of evolutionary development from inception to recent innovations.

The primitive beginnings of this plant-animal relationship are represented by beetles, thought to have been the pollinators of the first flowering plants. Clumsy and destructive compared to the more advanced pollinators, they feed mainly on pollen rather than nectar, visiting open, bowl-shaped flowers, and transferring some pollen in the process. Beetles are still very important pollinators in the tropical rain forest, although elsewhere in the world they have yielded their place to the more sophisticated nectar-feeding insects – bees, wasps, flies, butterflies and moths. But these more modern insect pollinators are found in abundance here as well, often serving flowers of astonishing complexity, such as the orchids. Some of these flowers rely on a single species of pollinator and are highly specialized to encourage that pollinator and exclude all other nectar eaters. The most extraordinary of these specific relationships involves figs and fig wasps, where the tiny wasps act out their entire life cycle within the developing fig, the newly hatched young leaving to fly directly to a new fig, and thus transferring the pollen.

The most recent recruits to pollination are the birds and bats, which can exist only in the tropics where there is a year-round supply of flowers. Over many millions of years, plants in a number of different families have switched allegiance from insects to vertebrates. They pay the price in extra nectar and larger flowers but in exchange they get a pollinator that can fly much further to seek out another plant of the same species. A few rain forest trees are even pollinated by lizards, and some eucalypt species in Australia are served by a marsupial mammal, the sugar glider (*Petaurus breviceps*).

As food nectar's only serious disadvantage is a lack of proteins. Birds and bats solve the problem by eating their fellow diners, it being a simple matter for a hummingbird or flower bat to collect up the insects at a flower while also taking its nectar. The insects themselves may eat pollen, or they may sip at corpses, dung or rotting fruit for their protein and mineral supplements.

Nectar thieves, which take nectar without transporting any pollen, are a major problem that plants have to contend with. As with seed thieves (see pages 76–79) this is not a situation that the plant can ever win, but plants gradually evolve defences, which may in turn be thwarted, wholly or partly. Successful nectar thieves include the flower-piercers and some hummingbirds which peck through the base of the flower to get at the nectar, and monkeys such as tamarins and capuchins which eat the entire flower. On the other hand there are many obvious signs of a temporary victory by certain plants. These include flowers that are inaccessible to monkeys on the edge of a tree's canopy, and those with their nectar secluded at the end of a long tube, or a curved one, making it available only to pollinators with the right mouthparts. Chemical defences are less common but a few flowers spike their nectar with poisons.

Simply collecting the pollen is not enough – for pollination to happen it must also be transferred, so the pollinator must move from flower to flower. Rather than supplying nectar in a continuous stream, some rain forest trees, such as the bat-pollinated *Oroxylum* sp. of Malaysia, dole out their nectar in small amounts at each visit, forcing the bat to move on to another flower for more food.

Complete diet (*left*) – The ithomid butterfly feeds at one of the few plants that fortifies its nectar with amino acids, the building blocks of proteins. It is generally insect-pollinated flowers that come nearest to providing a balanced diet in their nectar.

A perfect fit (*main pic*) – As the bronzy hermit humming-bird (*Glaucis aenea*) sips nectar from a passion flower, the flower's anthers brush pollen onto the back of its head. The close fit between the bird's head and the flower's shape is the product of coevolution.

Mating swarms

Euglossine bees illustrate the complex web of plant-animal relationships in the tropical rain forest. These small jewel-like bees are some of the most vital pollinators in the Amazonian forests. They are the sole pollinators for many species of orchid, and important for a huge variety of other plants.

In some species of euglossine bee, the pollinated plant acts as a focus for the bees' breeding activities. By collecting scented chemicals from the orchid flower, male bees generate a cloud of perfume that attracts other males, who repeat the process and thus augment the scent. Eventually, the fragrance is strong enough to draw the females, and mating takes place. By visiting a number of different orchids for their perfume, the male bees carry pollen from one to another.

Female euglossine bees generally feed on nectar from completely different plants, and some pollinate the Brazil nut tree (*Bertholletia excelsa*). Attempts to grow Brazil nuts in plantations have usually failed because the orchids, and thus the bees, were absent.

Studies of forest fragments in Amazonia have shown that several species of euglossine bee will not cross cleared areas, even though these may be only 100 metres (330 feet) wide. A large island of forest in a cleared area, extending to 100 hectares (250 acres), will be depleted of several bee species, and a fragment of just one hectare (2.5 acres) will lose about a third of its species. Such losses mean that some plants are no longer pollinated, and will eventually die out.

Leaf eaters

Every square metre (11 square feet) of lowland rain forest carries enough leaves to cover up to 11 square metres (118 square feet), and each year it produces up to 0.8 kilogrammes (one pound twelve ounces) of new foliage. Leaves contain a good mix of nutrients, and are a far better source of protein than nectar or most fruits. More importantly, leaves are there all the year round. Despite these attractions, only about 25 percent of the year's new leaf growth goes to leaf eaters, because rain forest trees have evolved a highly sophisticated set of defences.

The main deterrent to a leaf eater is that leaves are difficult to digest. Their principal ingredient is cellulose, which makes up the plant's cell walls. Some bacteria can digest cellulose, and so can a few wood-boring insects and other invertebrates, but most animals cannot. For a small leaf eater such as a caterpillar the answer is to chew the leaves very thoroughly, breaking open the cells to release the nutritious cell contents. The cellulose is then excreted unchanged in most species.

Larger leaf eaters need to get more energy from their food, and cannot chew it so finely, so those that eat leaves exclusively must break down the cell wall as well. Specialist leaf eaters, such as howler monkeys, maintain a colony of suitable bacteria inside the gut to do this for them.

Bacteria take a little time to work their magic on the leaves, and progress through the gut is slow. This makes the leaf eaters relatively heavy and most are not particularly fast-moving. By way of compensation, their food supply is all around them and makes no attempt to escape. Not surprisingly, few birds are leaf eaters, because heaviness and flight are generally incompatible.

Overcoming the plant's defences

Indigestibility is just one line of defence against leaf eaters; other defences may be particularly fearsome in tropical rain forests, where leaf-eating insects are legion (see page 84). Plants deploy a variety of weapons, including prickles, thorns, spines and, above all, toxic chemicals. Most heavily defended are the trees of the

heath forest, where the infertility of the soil encourages the plants to conserve their leaves, and thus their nutrients, as much as possible.

A typical heath forest tree is the Brazilian rubber tree (*Hevea brasiliensis*), which oozes white milky latex from its trunk, branches and leaves if these are punctured. Contact with the air turns this liquid to a sticky gum that plays havoc with a leaf-eating insect's mouthparts. The latex also contains toxins – yet even this double defence can be overcome. Several insect species disarm the tree by cutting off the supply of latex to part of a leaf before eating it. This is achieved by punching a line of small holes in a leaf or by severing the main latex duct.

Several other rain forest trees use latex for defence, whereas others protect their leaves with straightforward toxins, some of which are useful to humans as medicinal drugs. But however powerful they may be, all have been overcome by one leaf eater or another. Leaf-eating insects – which are probably the target of most chemical defences – have evolved a battery of powerful enzymes to break down specific poisons. This generally limits them to certain related groups of plants which are chemically similar. From the plant's point of view, the chemical defences do not exclude all leaf-eating insects, but they limit the amount of damage done. Although the chemical barriers are erected primarily against insects, they affect other animals as well. Monkeys and other animals with cellulose-digesting bacteria in the intestine get help with detoxification from these microorganisms, so they too are immune to some poisons.

Having acquired immunity to these poisons, the leaf eater may use them as food, or store them in its body as a protection against its own enemies. The caterpillars of the huge birdwing butterflies of New Guinea feed on poisonous lianas belonging to the genus *Aristolochia*. They store the plant's toxins and pass them on to the adult butterfly, whose striking coloration is a warning to potential predators of the poisons within.

The stinkbird – One of the few leaf-eating forest birds is the hoatzin (*Opisthocomus hoazin*) of South America. To help with digestion, it has bacteria in its greatly enlarged oesophagus and crop. This is equivalent, for a bird, to the stomach fermentation of cows, which explains why the hoatzin (also known as the "stinkbird") smells like cow manure. Leaves stay in the hoatzin's stomach for almost 2 days, and the overloaded bird is in consequence a very poor flier.

Fungus farmers (*left*) – Leaf-cutter ants solve the problem of extracting the goodness from leaves by enlisting the help of a fungus. They first cut up the leaves, using their scissor-like mandibles, into easily transported sections, and carry them back to their underground nest. The leaves are then chewed into a pulp and the spores of a particular fungus added. The fungus grows on the pulp, extracting the nutrients, and the ants feed on the fungus.

Tough diet (*above*) – The proboscis monkey (*Nasalis larvatus*) lives in flooded forests, mangroves, and heath forests in Borneo, feeding on tough leaves and seeds. In its large stomach there is an array of bacteria which help with the monkey's digestion by breaking down the cellulose in the leaves and detoxifying any plant poisons that are present.

Chewing for sap – Trees make sugar in their leaves (see p. 61) and send this down to their trunk and roots to keep these parts alive and growing. The sugars travel down the trunk in the form of a sweet liquid, known as sap. Several mammals in the rain forest tap into this food resource, notably the South American pygmy marmoset (*Cebuella pygmaea*), pictured here, and the needle-clawed bush baby (*Euoticus elegantulus*) of Africa. For both these species it is the main element in the diet, whereas other small monkeys and prosimians (bush babies, pottos and lemurs) eat sap occasionally. Pygmy marmosets chew out holes from the trunks of trees, and visit the trees at intervals to lap up the sap that gathers in the holes. A family group has a small home range within which it taps all the suitable trees, moving on to a new area of forest about once a year when all the trees have been exploited to the full. Trees guard their trunks and branches as jealously as their leaves, and many produce clear sticky gum to congeal the mouthparts of wood-boring insects. Far from being deterred by gums, sap-feeding mammals generally turn this habit to their own advantage and feed on the gums as well. They are less digestible than sap, and gum feeders may be aided in breaking down their food by bacteria living in the gut. Mammals that feed on sap and gum have various adaptations. Sharp, curved claws allow them to cling to tree trunks and move vertically up or down them. The marmosets have large front (incisor) teeth for gouging out holes in tree trunks. The enamel is thin on the side nearest the tongue, and soon wears away, leaving a chisel-like tooth which is very effective at this task.

Predators

Compared with the plant eaters of the rain forest, the predators are far less diverse, at least in terms of numbers of species. Yet they are still more varied and numerous than in other types of forest. The insect eaters, which have a wealth of invertebrate life to feed on, are particularly prevalent: the main invertebrate predators are in fact invertebrates themselves. The flesh eaters of the forest are notable mainly for their small size, there being too few large herbivores to sustain a large carnivore. The exceptions are the large cats, such as the jaguars of the New World and the forest tigers of the Old World. Confined to the forest floor and lower branches, they prey on animals such as deer, tapirs, peccaries and other wild forest pigs.

Insect eaters

To feed on insects, an animal must first overcome their often formidable defences, and rain forest insectivores tend to specialize in the type of prey they capture. The defensive strategies of insects fall into four main groups: concealment, camouflage, making a rapid escape, or being poisonous, prickly or aggressive. A fifth, and more devious, strategy is mimicry, in which a palatable species resembles a poisonous one.

Searching for insects that defend themselves by concealment provides a good living for birds such as caciques and oropendolas, found in the rain forests of the New World. These all have well-developed muscles in the head enabling them to open their bills forcibly against external pressure. They use this technique to prise apart the leaves of bromeliads and other epiphytes, revealing insects within. Saddle-back tamarins may be found in the lower storeys of the same forests, exploiting a different group of hidden insects, those in tree holes and bark crevices. The tamarins' ability to cling on to vertical trunks brings this otherwise inaccessible food source within reach. In the forests of Asia, spiderhunters use long curved bills (half as long as their bodies in some species) to probe branches, flowers and vegetation for insects.

Hidden and camouflaged insects are vulnerable the moment they move, and many rain forest insectivores rely on disturbing such insects to capture them. The large mixed flocks of insectivorous birds that move through the canopy together (sometimes called "bird waves") do just this, each eating the insects that its neighbours' feet have disturbed. Army ants and driver ants (see page 86) achieve the same effect on the forest floor.

Those insects that rely on speed to escape predators may also fall victim to birds foraging in the canopy in large flocks. One bird of prey, the double-toothed kite (*Harpagus bidentatus*) of South and Central America, specializes in prey of this type, relying on troops of squirrel monkeys to drive them out.

Other predators lie in wait for insects of this type, including the forest-dwelling chameleons and the flower mantids. For the night-flying insects such as moths, the large numbers of insect-eating bats that patrol the rain forest are the greatest threat. Some catching their prey above the canopy, others feeding within the forest.

Insects that defend themselves with bites, stings, spines and poisons are more difficult to eat, but not impossible. The lorises and pottos of the African rain forests specialize in this type of prey, and are apparently resistant to their chemical defences. Caterpillars covered with irritant hairs can cause an intensely itchy rash on human skin, even if brushed only lightly. The golden potto (*Arctocebus calabarensis*), also known as the *angwantibo*, eats such caterpillars, rubbing them between its hands first, and afterwards wiping its lips and nose against a branch. Studies of captive pottos show that they are not totally indifferent to the toxins in their prey, because they will select more palatable insects if offered a choice. By adapting to prey that does not attempt to escape, the pottos and lorises can afford to be slow-moving themselves.

Chemical deterrents and painful bites are typical of ants and termites, whose aggressive defence of their colonies deters most predators. The main predators of termites are ants, but the

Double-jointed hawk – The crane hawk (*Geranospiza caerulescens*) is a predator of swampy forests in Latin America. It specializes in extracting prey from tree holes and crevices. The inter-tarsal or "ankle" joint, which in most birds can move in only one plane, is far more flexible in this hawk. It can allow the lower part of the leg to bend backwards as well as forwards, so the hawk can thoroughly probe the inside of tree hollows while hanging on to the trunk with its other leg. The hawk's legs are also unusually long, and its head is very narrow, to reach inside nooks and crannies. Crane hawks prey on tree frogs, smaller birds and their nestlings, eggs, lizards and larger insects. A very similar species, the African harrier hawk (*Polyboroides typus*), takes the same type of prey in the forests and woodlands of Africa. It is not known whether the two are related, or have simply developed shared attributes through following the same way of life.

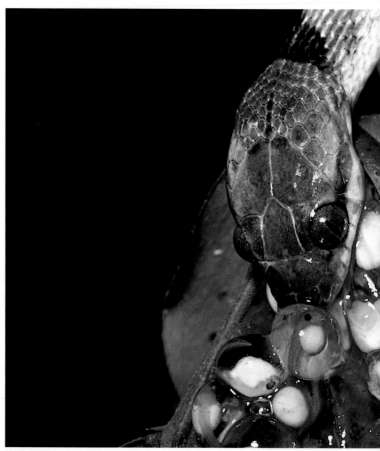

Time to get away (*above*) – Some camouflaged insects such as this eyed silkmoth (*Automeris rubrescens*) have a second line of defence, using startle displays to alarm a predator that has disturbed them. Such tactics are most effective against birds, and the display often consists of highly realistic eye-like markings that are revealed suddenly when the insect is discovered. By resembling the eyes of a predatory civet or snake, the moth can startle the prey momentarily, giving itself enough time to escape. Although eye-like markings are not exclusive to the rain forest they are particularly elaborate here.

Frog predator (*left*) – The cat-eyed snake (*Leptodeira septentrionalis*) is a highly specialized predator whose diet is only possible in the rain forest. The bulk of its food comes from tree frogs and the eggs which many species, including the glass frogs and leaf-folding frogs, lay on the leaves of rain forest trees. Although they may try to defend their eggs by guarding them or by wrapping leaves around them, the frogs are no match for this canopy predator.

Blood suckers (*above*) – Animals that drink blood can be classified as predators or parasites – they lie somewhere between the two. Among the largest are the vampire bats, the Desmodontidae, of which there are 3 species found only in the American tropics. They do not suck blood, but remove a small piece of skin and then lap up the blood flowing from the wound. An anticoagulant in the bat's saliva prevents the blood from clotting for as long as the bat is still feeding. Recent research has revealed that when vampire bats return to their roosting sites at the end of the night they may regurgitate blood for close companions who have not found food. This form of sharing is uncommon among mammals.

85

colonies contain so much food they also attract larger, specialist feeders. In the rain forests of the New World, the tree anteater (*Tamandua mexicana*) and pygmy anteater (*Cyclopes didactylus*) raid the nests of ants and termites high up in the trees; in the African and Asian forests, tree-climbing pangolins exploit the same food source. They are equipped to repel the ferocious attacks of their prey, with thick fur or bony scales (see page 48).

Ant and termite nests attract other less specialized predators, such as capuchin monkeys which break them open with their strong hands. Unlike anteaters, they do not relish the adults, but lap up the eggs and larvae. Their thick fur apparently defends them from the assaults of the colonies' soldier castes. In Southeast Asia, a great many ants and termites are taken by woodpeckers, which can peck the nests open with their strong beaks, and by game birds, which scratch them out of the soil.

Nutritionally, insects are rich in protein and minerals, and many animals use them to supplement diets of fruit or nectar, which are lacking in these components. Nectar-feeding birds such as hummingbirds and sunbirds raise their nestlings on young insects to promote rapid growth. For some omnivorous animals, such as squirrel monkeys, insects are not part of their diet all year round but become a source of emergency food in times when fruit is scarce. However, a squirrel monkey is too large to live entirely on insects for more than a few months. It requires more energy just to stay alive than a small insectivore such as a tamarin, but it cannot catch prey any faster than the tamarin. This same principle affects all the large monkeys and apes. Above a certain body size they cannot afford the energy needed to catch insects, even as a protein supplement, and must turn to leaves instead.

Flesh eaters
Some of the most successful flesh eaters of the rain forest are the hawks and eagles, which can spot prey from a vantage point above the forest while on the wing or perched in an emergent tree, then swoop down into the canopy to seize the prey in their talons. These airborne hunters take a variety of smaller birds and mammals, the largest eagles specializing in such prey as monkeys, lemurs and sloths.

For most other rain forest predators, moving in for the kill is far more problematic, and climbing skills are at a premium. Snakes do well in the canopy, as do small, agile members of the cat family, such as the margay (*Felis wiedii*) of South America. African and Asian forests are home to many civets, linsangs and genets, a family of slender, elongated animals with long tails and short legs. Such a body shape is ideal for balancing on branches, and many are almost entirely carnivorous, taking lizards, rodents and frogs, as well as many worms and ants. On Madagascar, where there are no big cats, a large predatory species, known as the fossa (*Cryptoprocta ferox*), has evolved from this group. The size of a large, sturdily built terrier, with a foxy face and a very long smooth tail, it preys on lemurs, birds, lizards and snakes.

One of the effects of the constant warmth of the tropical rain forest is that there are fewer restrictions on the size of insects and other invertebrates. Species have evolved which can grow to extraordinary sizes by temperate standards (see page 53). Some of these giant invertebrates have turned the tables on the vertebrates and include them in their diet. Tropical mantids can grow up to 20 centimetres (eight inches) long, and some prey on tree-dwelling lizards, young birds, frogs or even small mammals. Some of the large mygalomorph spiders of South American forests (often called tarantulas) are tree-dwellers which catch finches and other small birds, probably while they are sitting on their nests. Other spiders in the same group take lizards, frogs and even small poisonous snakes, pouncing on the head of the snake and delivering a venomous bite. Spiders can only consume liquid food, and these large prey items are macerated with the powerful mouthparts and sucked dry, a process taking an hour or more.

Efficient predators – Nomadic army ants live by mass-predation, flushing insects from the leaf litter as they advance through the forests, or mounting raids on the nests of other ants, termites or wasps. At night the worker ants form a temporary "nest" by surrounding the queen and her larvae with their bodies (*right*). This unusual way of life has evolved on at least two separate occasions, in Africa (where some are known as driver ants) and in the American tropics. So effective is the ant's feeding techique that it leaves the undergrowth largely devoid of life for several weeks or months, and it is this that has forced the army ants to become nomadic. An added advantage of this technique is that it allows the ants to overcome quite large insects which they would not be able to tackle individually. Even scorpions, small snakes, lizards and nestling birds may fall victim to the marauding ant column. Many birds benefit by picking up insects which are fleeing for their lives from the advancing ants. Other insectivores have occasionally been seen following army ants, including toads, lizards and marmoset monkeys.

Canopy hunter (*above*) – The green tree python (*Chondro-python viridis*) from New Guinea is a predator of tree frogs and an example of convergent evolution with the emerald tree boa (*Corallus caninus*) of South America. The two look very similar: both have prehensile tails and both coil themselves around a branch when at rest. Yet these two snakes are not at all closely related. It is only the force of natural selection, in their shared habitat and way of life, that has made them similar.

Outsized spiders (*left*) – The large, wandering spider *Cupiennius coccineus* preys not only on other invertebrates but also vertebrates such as tree frogs. The warmth of the rain forest allows cold-blooded invertebrates to exceed their usual size. The wandering spiders, which are found only in the tropics, resemble the wolf spiders of temperate regions. Like wolf spiders they are hunters.

Jaguar (*above*) – The jaguar (*Panthera onca*) feeds on monkeys, capybaras, deer, birds and lizards, but its preferred prey is peccaries, the wild pigs of the Americas. Jaguars hunt in rivers as well as the forest, taking fish, otters and turtles, and after the floods recede they may scavenge for stranded fish and alligators. In Surinam, they even prey on sea turtles on their nesting beaches. But jaguars do not generally leave forested areas, and despite the lure of the peccary, they do not move back into felled areas. Hunting has made the jaguar rare.

Tree travel

A schoolchild, putting his hand up in class, holds his arm flat against the side of his head. It is a movement impossible for most mammals, and a reminder that human ancestors once lived in the tropical forest. The anatomical arrangement that allows this action originally evolved for swinging from branch to branch by the arms. The most skilful modern practitioners of this form of travel are the gibbons and siamangs, which swing through the canopy of Asian rain forests at breathtaking speed.

These fast-moving apes represent one strategy for travel through the forest. It is a high-risk, high-return strategy: falls occur and can result in broken bones, but the rewards are trees loaded with ripe fruit. By moving rapidly through the canopy, the foraging gibbons can locate these scattered resources. In the New World, spider monkeys pursue the same strategy, racing through the canopy at such speed that an observer on the ground has difficulty in keeping pace. Less specialized for swinging, the spider monkey jumps, swings or runs along horizontal branches, constantly changing its gait. When necessary, it can bring its tail into play, winding it tightly around a branch to act as a "fifth limb". This prehensile tail proves even more useful during feeding, especially among slender, fragile branches, where the monkey can use its tail to spread its weight more widely and avoid crashing to the forest floor.

At the other extreme are the sloths, adapted for a sluggish low-energy lifestyle, fuelled by a diet of easily found but indigestible leaves. Sloths are so thoroughly adapted for hanging upside down that no effort is involved, and they remain in position even when dead, suspended by long curved claws and decomposing slowly in mid-air. Extreme slowness is also seen among the lorises and pottos of Africa and Asia. Like the sloth, these creatures rely on stealth to evade predators, moving cautiously and freezing at any unfamiliar noise or movement. To the frustration of biologists who have tried to study them, they can remain motionless for hours if alarmed. This strategy only works well in thick cover, and lorises and pottos are confined to the luxuriant vegetation of the rain forest.

A third approach to tree travel is to cling to the trunks of trees, leaping from one to the next. Substantial force must be generated by the hind-legs, and this form of movement is found mainly among smaller forest dwellers such as the tarsiers of Southeast Asia and the bush babies of Africa. Neither of these groups occurs in South America, and here the tiny monkeys known as marmosets (see page 83) and tamarins fill the same role. Unlike other monkeys they have claws, rather than nails, for clinging to the tree trunks. Their strategy fails if the tree trunks are too far apart, and they cannot inhabit the tallest forests, but do well in swamp forest where there is a thicket of small stems.

Flying through the forest

A leaping animal can go further if it can somehow defy gravity and fall more slowly. This can be achieved with flaps of skin that act like parachutes, enlarging the surface area of the animal so that it drifts gently down (see right). Gliding animals include flying squirrels, flying lizards, flying snakes, flying frogs and, most accomplished of all, the colugo (*Cynocephalus variegatus*).

None of these animals truly fly, despite their names. That is left to the birds, bats and insects that inhabit the tropical rain forests in great numbers and enjoy the benefits of easy travel between its different storeys. Only the dense foliage of the canopy presents problems, making navigation difficult during flight, especially for the bats who rely on a system of echo-location, akin to radar. Larger bats find it easier to fly beneath the canopy, and the trees that rely on them for pollination or seed dispersal produce their fruit and flowers on their trunks. Those bats that do venture through the canopy use habitual flight paths to avoid dangerous obstacles such as branches obscured by leaves. Likewise, insects follow set routes as they fly through the canopy, and many spiders take advantage of this predictable behaviour when positioning their webs.

Ungainly glider – The flying gecko (*Ptychozoon kuhli*) uses two flaps of skin, one each side of its body, to glide through the canopy. These flaps are wrapped around its belly until it dives off a branch; then air pressure forces them open, converting the gecko into a rather inefficient glider. The gecko also uses its flattened tail and webbed feet to control its fall.

The ribbed flyer (*above*) – In the rain forests of Southeast Asia, where hilly terrain and tall emergent dipterocarp trees make gliding especially useful, the flying dragon (*Draco volans*) uses a loose membrane of skin attached to its extended ribs to swoop from tree to tree. Before leaping into the air, it extends its ribs, forming a stiff "wing" on each side of its body.

Marsupial in flight (*top right*) – The Australian sugar glider (*Petaurus breviceps*) is an excellent nocturnal flyer. It has large folds of skin between its legs which it stretches taut to form a furred "parachute".

Flying frog (*right*) – The South American frog *Agalychnis spurrelli* relies on the webbing between its toes to slow its descent. It is capable of short glides of 12 m (40 ft), and by extending or retracting its limbs it can steer to the right or left.

The fifth limb (*above*) – Some agile New World monkeys, like this black spider monkey (*Ateles paniscus*), use their prehensile tails to help them move quickly through the trees. Other slower-moving species, such as the tamandua anteater and tree pangolin (see p. 53), use their prehensile tails for stability.

People of the rain forest

In the remote rain forests of Borneo, fragments of medieval Chinese pottery have been found, and naturalized fruit trees that can only have been brought there, perhaps as seeds, by human hands. In parts of Amazonia, unusually fertile patches of black soil are sometimes discovered. Archaeologists have shown that these oases of nutrient-rich soil were created by adding large amounts of organic matter for intensive cultivation. These and other tantalizing fragments of evidence show that people have inhabited the rain forests for many thousands of years, and lived in a great variety of ways, only some of which have survived to the present day.

No one knows where most of the rain forest people came from, how they colonized the forest, or how they are related to each other. Warmth and moisture conspire to break down all organic materials with great speed, so ancient remains are rare, even where people buried their dead. Many forest tribes had no knowledge of metal, and no need of it, since there were ultra-hard forest woods that could make just as deadly an arrowhead as any iron or steel. But wood, however hard, rots down over the years, and so do all the other natural materials that rain forest peoples use: the bamboos and rattans, the liana ropes, plant-fibre fishing nets and hammocks, the wooden buildings and woven baskets. Generations of inhabitants lived and died leaving scarcely a footprint on the forest.

In Amazonia, only pottery, which first appeared about 3,500 years ago, provides any clues about early inhabitants, and by this time people had already been in the region for at least 9,000 years. Linguists can trace the lines of migration by looking at relationships between the languages of present-day tribes, and anthropologists can compare physical traits and customs; but much of what happened in the past remains guesswork.

The first people to make a living from the rain forest were probably hunter-gatherers, who survived by fishing, catching game and gathering plant foods such as fruits, nuts and tubers. Agriculture may have been brought into the forest by newcomers from elsewhere, or developed independently by the original forest inhabitants. In Central Africa, it is clear that agriculture was introduced from outside as immigrant Bantu tribes moved southwards into the rain forests during the Bantu expansion of about 2,500 years ago. The impetus for this may have come from metal-working: the Bantu had acquired a knowledge of iron-smelting through their trade contacts, and this made it easier to clear the rain forest for cultivation. Prior to that, Pygmy hunter-gatherers were the only human inhabitants of the forest.

In Malaysia, too, farming methods were probably introduced from outside, since most of the present-day agriculturalists belong to different racial groups from the hunter-gatherers, being taller and fairer-skinned. In Amazonia there may have been an influx of agriculturalists about 2,000 years ago, as shown by the appearance of a new type of pottery, decorated with bold geometric patterns. These newcomers developed semi-permanent settlements based on the highly productive *várzeas*, or flooded areas, where whitewater rivers deposit nutrient-rich sediments. But these were probably not the first cultivators to live in Amazonia, and the distinction between hunter-gatherers and agriculturalists is not as clear-cut here as it is in Africa. Many tribes are shifting or "slash-and-burn" cultivators, who also gather and hunt wild foods. The importance of crops in their diet varies greatly. Some are primarily hunter-gatherers who also grow a little food, while others are cultivators who supplement their diet by hunting or fishing.

Xingú hunt – The lifestyle of the people who live around the Xingú River, a tributary of the Amazon, has over the past two decades been severely disrupted by outside interference.

Undisturbed, they spend only about 4 hours a day searching for the food they need.

Hunter-gatherers

Human beings are not creatures of the forest. We evolved on the open savannas of Africa, where big game was plentiful and the hunting easy. Our earliest ancestors did not even hunt, but lived the same sort of life as vultures, scavenging on the carcasses of large animals that had been killed by predators or simply died a natural death.

On the savanna, about 50 percent of all the new vegetation is eaten by the herds of antelope, zebra and other large herbivores. By contrast, only 2.5 percent of the new plant growth in a rain forest is used in this way. The savanna-type prey of humans is very scarce in the forest, and in its place are countless insects, birds, reptiles, monkeys, bats and rodents. Not only are these forest animals small, but they are elusive and inaccessible in the lofty canopy. It took thousands of years for humanity to conquer this demanding habitat.

Partly because there is relatively little game to be had, population densities in the forest have never been high. In the Zaire Basin, the population density of the Mbuti Pygmies is only about one person for every four square kilometres (1.5 square miles). An individual band may range over a vast territory in its search for food, as much as 1,300 square kilometres (500 square miles), but famine is unknown – the forest always provides.

The diet is not only plentiful but also very varied. Indeed, the ability to utilize a wide range of foods is a key to survival. The Semang of Malaysia take nuts, berries and other fruit, young leaves and shoots, roots and tubers, honey from wild bees, fish, birds, rats, squirrels, lizards and occasionally wild pigs, tapirs and deer. They eat at least ten different species of wild yams, and a wide variety of fruits including the durian, rambutan and tampoi. Some foods, however, are taboo, notably leeches, scorpions and spiders. The Amazonian tribes have a similarly varied diet: to the Siriono of Bolivia, the only taboo foods are insects. For other groups, such as the African Pygmies, fat beetle grubs and caterpillars are an important source of protein at certain times of the year.

Honey is important to most of these hunter-gatherers, despite the difficulties of collecting it from high in the trees. Many tribes use smoke to drive the bees away or stupefy them, before approaching the nest. The Baka Pygmies scale the tree using a rope harness made of lianas, and a home-made wooden tool that resembles a mountaineer's ice-axe, with which the climber can hack into the trunk and then pull himself upwards.

To capture monkeys, hornbills, parrots and other prey high up in the canopy, rain forest hunter-gatherers use poisoned arrows or darts. It is an ingenious technique which was invented independently in different parts of the world. The sources of the poisons are many and varied. A wide range of different plants are employed, and a few Amazonian tribes use the skin of toxic tree frogs. The poisons used are mostly inactive if taken by mouth, so the meat is good to eat. The Semang of Malaysia have a special weapon for delivering poison – a dart propelled from a blowpipe. Only one type of bamboo, *Bambusa wrayi*, produces the raw material for these blowpipes, which call for straight, two-metre (six-foot) long sections of stem. Every clump of this bamboo species is known, and access strictly limited by those in whose territory it lies.

Living a nomadic existence in small family groups leads to a relatively relaxed social order. Most of these hunter-gatherers are egalitarian and non-hierarchical. Where there are chiefs – as in the Amazonian tribes – they usually have little power to command others against their will. A leader must earn the respect and cooperation of his group through his own personality and competence. Among the Baka Pygmies, there are no chiefs, and social harmony is achieved by consent and compromise. At the close of each day, after story-telling, joking and gossiping, one of the old men or women in the group will invariably deliver a formal "advice-speech" to the band, commenting on events, suggesting ways in which differences might be settled, and reinforcing traditional values and beliefs.

Adaptations to the forest

For human beings, life in the rain forest requires, above all, cultural adaptations. The ability to identify thousands of different types of plants and animals, in a biological kaleidoscope that expert field biologists find daunting, is the secret of success. Forest tribes must remember the fruiting cycles of forest plants, recognize hazardous insects and reptiles, and be able to hunt elusive prey. Social constraints which restrict population growth may also have helped develop a sustainable way of life.

Against the background of these cultural adaptations, physical changes are relatively unimportant, but these too play a part. Most noticeable is the small stature seen among the Pygmies of Africa and the so-called "Negritos" of Asia: the Semang of Malaysia, the Onge people of the Andaman Islands and the Aeta of the northern Philippines. These groups almost certainly represent ancient forest inhabitants, who have lived in the rain forests for so many thousands of years that they have had time to evolve physical characteristics suiting them to this environment. Among animals, there are many forest species that are substantially smaller than their counterparts elsewhere – for example, the pygmy hippopotamus (*Choeropsis liberiensis*), the forest elephants of Africa (*Loxodonta africana cyclotis*), and the tiny royal antelope (*Neotragus pygmaeus*). In dense undergrowth, smaller stature may make it easier to move about; and for humans, a light, muscular frame is better suited to tree-climbing. These small, lightweight peoples are traditionally hunter-gatherers although many have now relinquished their old way of life. Human populations that have moved into the rain forests more recently may show little reduction in size. Amazonian tribes are generally smaller than lowland Amerindians elsewhere, but the difference is not great.

Hunting (*left*) – In northern Zaire, the Mbenga Pygmies obtain a lot of their food by hunting; the Bantu- and Sudanic-speaking peoples (of whom there are many times the number of Pygmies) rely on shifting or permanent cultivation, and on herds of domesticated cows, sheep and goats.

Gathering (*above*) – Unfortunately, bees that sting generally produce more honey than bees that don't, so to obtain such a highly-prized addition to the diet many forest peoples use smoke to drive away the insects before attempting to collect the honey.

93

Agriculturalists

The Siriono of northern Bolivia are an intriguing Amazonian tribe who can only be described as "nomadic agriculturalists". They get most of their food from hunting and gathering, but they also clear garden areas in the forest at certain times of the year. They plant crops, and then move on, allowing the fruit and vegetables they have sown to battle it out with the forest vegetation unaided. The Siriono hunting band continues its wanderings through the forest, only returning to the garden several months later to gather the harvest. Once the crop has been harvested, the plot is abandoned and reverts to forest. This is shifting cultivation at its very simplest, feasible only where wild foods provide the major part of the diet.

Though few are as dependent on wild foods as the Siriono, all rain forest agriculturalists engage in some hunting or fishing, and most gather fruit, nuts, honey, fungi and other delicacies from the forest. In a sense, shifting agriculture is a logical extension of gathering plant-food direct from the forest, in which the forest is encouraged to produce more of its useful vegetation (together with some crops brought in from elsewhere) in a convenient and accessible spot. Indeed, the forest gardens of shifting cultivators have a remarkably natural look, often mimicking the layered structure of the surrounding forest. This is how one anthropologist described such gardens in New Guinea: "To enter the gardens is to wade into a green sea. To walk is to push through irregular waves of taro and xanthosoma (ferns) and to step calf-deep in the cover of sweet potato vines. Overhead, bananas and sugar cane provide scattered shade. . . ." The range of crops grown is staggering. One survey found that the Temuan in Malaysia exploit 65 species of fruit tree, both wild and cultivated. Shifting cultivators rarely keep domesticated animals, relying on wild creatures for their protein. And they are essentially nomadic, as hunter-gatherers are, although a few tribes manage to stay in the same place for several generations by using a succession of plots within a large radius of their village. On average, shifting cultivators change their living site after 15 to 20 years. The Iban of Sarawak, an exceptionally mobile group, move by 90–180 kilometres (50–100 miles) each generation, and sometimes as much as 460 kilometres (250 miles) in a lifetime.

For those cultivators who stay in one place, tending permanent fields and plantations, there is a greater sense of detachment from the forest. Domestic animals are less of a rarity, and the cultivated plots are more estranged from forest vegetation. Whereas shifting cultivators have a temporary impact on the forest, allowing it to revert to its natural state afterwards, settled cultivators remove the forest cover permanently and replace it with an essentially artificial plant community. But even these cultivators rely on the remaining forest for a substantial amount of their food, or as a source of income. The Temuan of Malaysia are horticulturalists, but they gather rattans from the forest to sell or barter. Often, as with the Bantu tribes of Central Africa, settled cultivators obtain forest produce though trading relationships with hunter-gatherers – in this case Pygmies – who receive crops or manufactured goods in return.

Permanent cultivation is only possible in certain rain forest areas, where the soil is sufficiently fertile and other conditions are favourable. River plains tend to provide such a setting, the sediment brought down by the river renewing the fertility of the fields each year. In Amazonia, there was formerly a thriving Amerindian farming community based on the Amazon's flood plains or *várzeas* (see pages 128–129). In West Africa a relatively fertile soil allowed the development of thriving agricultural economies, such as that of the Asante, based on subsistence farming combined with cocoa and other plantation crops.

In social terms, the agriculturalists live quite different lives from the hunter-gatherers. Permanent cultivators, in particular, live in larger groups that are far more complex and hierarchical. Shifting cultivators are less dominated by chiefs, but are rarely as egalitarian and easy-going as the hunter-gatherers.

The Yanomami use digging sticks to plant seeds.

Shifting cultivators in Colombia.

How shifting cultivation works

Shifting cultivation – also known as "swidden agriculture" or "slash-and-burn" – is a farming method well-suited to the impoverished soils of the rain forest. By burning the vegetation, shifting cultivators release nutrients that are locked up in the plants and enrich the soil temporarily, so that crops can be grown. The work on a garden plot begins long before that, however. First the site must be chosen. Some rain forest tribes select areas with a particular type of tree or other vegetation, because they know these will have relatively fertile soil. Once the plot has been earmarked, the fruits and other useful products found there may be harvested, often over a period of several months. Large trees may be removed to make houses or dug-out canoes. The undergrowth is then cleared and the remaining large trees felled. To avoid the massive base of the trunk, they are often felled fairly high up, by a man standing on a makeshift platform. Such stumps may survive burning and sprout again in time, which speeds the regeneration of the forest vegetation once the plot is abandoned. Some trees are left unfelled, either because they are useful or are regarded as sacred. After felling, some tribes remove timber to make fences. This is common in New Guinea where pigs, both wild and domesticated, can destroy crops.

Next comes the burning of the remaining foliage and timber, once it has dried out enough to ignite. Burning creates nutrient-rich ash to fertilize the soil, but, just as importantly, it eliminates many weeds. After burning, planting can take place straight away. Digging over the plot is unnecessary, because forest soil is already soft and friable, and most shifting cultivators use a simple digging stick to make a hole for their seeds. The charred trunks of fallen trees may still criss-cross the plot, but the crops are simply slotted in between them. They help to keep animals off the newly emerging plants and prevent the rain from washing away too much of the ash.

Once the crops are established, weeding has to begin, and this is a very time-consuming and arduous process. Many shifting cultivators limit the amount of land they farm because of the labour of weeding a large area. Weed invasion becomes worse with each successive growing season, and for many cultivators it is the rising tide of weeds that eventually forces them to abandon the garden and move on. In areas with very poor soil, such as the *terra firme* forests of Amazonia, declining fertility is more likely to spell the end of a particular garden plot. In either case, one to three years is the usual lifespan of a garden, and the plot will then be left fallow for between eight and 20 years. Shifting cultivation yields good crops for a modest input of labour, but it needs large areas of forest to support a relatively small number of people.

Atlas of the rain forests

The degradation and destruction of tropical rain forests is one of the most important issues of our time. Throughout the world conservationists, foresters, politicians and countless concerned citizens are joining forces to find ways of preventing further loss. But this international movement has been hampered by a lack of reliable data on the location and condition of the remaining forests. Surprisingly, no organization has ever published more than a sketch map to show the global distribution of rain forests. The maps on these and the next few pages represent the first attempt to bring together into a comprehensive whole the scattered information available from around the world.

On the world map (*right*) it might seem that plenty of rain forest remains. In fact, rain forest covers no more than 8.5 million square kilometres (3.3 million square miles) or six percent of the earth's land surface. This is just over half of what was not long ago a forested area of about 14 million square kilometres (5.5 million square miles). In 1980, the rate of deforestation of rain and monsoon forest was around 71,000 square kilometres (27,000 square miles) a year, about 0.6 percent of the remaining area, according to the Food and Agriculture Organization (FAO). But environmentalist Norman Myers has recently completed a study for Friends of the Earth (FOE) in which he estimates the current rate of loss to be a staggering 142,000 square kilometres (54,000 square miles) a year.

It was once thought that rain forest lost in this way was gone forever. It is now known that with care and sufficient respite most logged forests can be brought back into sustainable production for timber and other forest products, and can even provide habitats for wildlife. This book therefore takes a slightly more optimistic view than that suggested by the figures above and includes on the maps some forests degraded – but not destroyed – by logging.

When is a rain forest not a rain forest?

The lines demarcating rain forest from former rain forest on the maps are based on the best information available; but, for a number of reasons, these boundaries may not be as clear-cut on the ground as the maps would suggest. The relevant convention on such boundaries decrees that the canopy of what constitutes a genuine rain forest must cover more than 40 percent of the ground. Assessing this percentage is not easy, and in any case there is by no means universal agreement on the principle.

In addition to this, the boundary between rain forest and other types of tropical forest is equally difficult to judge, particularly in view of the fact that disturbance or actual damage by fire can cause the rain forests to degrade into these other forest types. For example, in Indochina rain forests degrade into drier monsoon forests in which annual fires tend gradually to open up the canopy, reducing the forest to open woodland with grass beneath. Further degradation leaves only scrub and bamboo behind.

It is not only on the forest edge that deforestation pushes relentlessly forward. Deep inside the forest, agriculturalists are clearing areas too small to show on the maps, and logging companies are exploiting the forest over thousands of square kilometres. By the time the first of such damage becomes apparent much of the whole of the forest may be degraded. But the rain forest does have the potential to recover, given enough time to do so. Even in Kalimantan and Sabah, where in 1982–83 drought followed by massive fires destroyed more than 40,000 square kilometres (15,500 square miles) of rain forest, reports are now emerging of extensive regrowth in the affected areas.

The vitality and robustness of the rain forests nonetheless permit no complacency. In Southeast Asia, the rain forests cleared for shifting cultivation are often left fallow for too little time for the soil to replenish itself. The unrelenting pressure of logging, relogging, clearing and burning that is taking place in much of Amazonia, West Africa and Asia is pushing the rain forests into a downward spiral of degradation from which there is little chance of recovery.

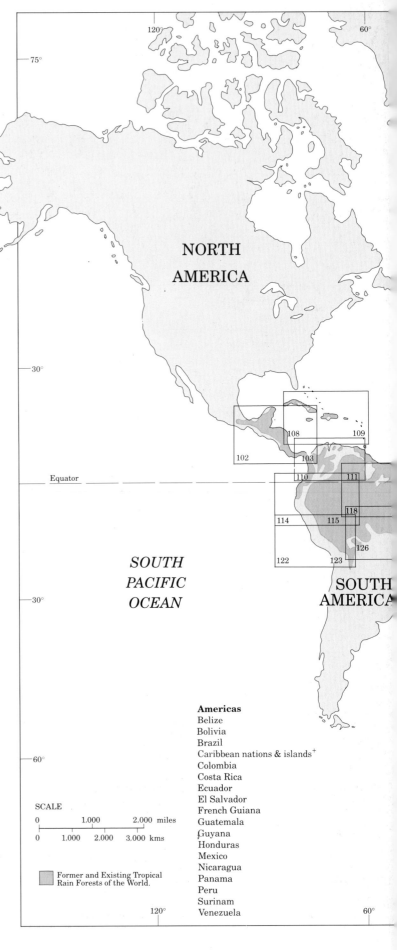

SCALE

Former and Existing Tropical Rain Forests of the World.

Americas
Belize
Bolivia
Brazil
Caribbean nations & islands[+]
Colombia
Costa Rica
Ecuador
El Salvador
French Guiana
Guatemala
Guyana
Honduras
Mexico
Nicaragua
Panama
Peru
Surinam
Venezuela

ARCTIC
OCEAN

60°   120°   75°

NORTH
ATLANTIC
OCEAN

E U R A S I A   60°

AFRICA   NORTH
PACIFIC
OCEAN   30°

155
156   157
150   163
136   137   161   166   167
140   141   170   171
145
INDIAN   174
OCEAN   AUSTRALIA   30°
146

SOUTH
ATLANTIC
OCEAN

Rain Forest Countries

Africa

Angola *	Madagascar	
Benin	Malawi *	
Burundi	Mauritius *	
Cameroon	Mozambique *	
Central African Republic	Nigeria	
Comoros	Rwanda	
Congo	São Tomé and Príncipe	
Côte d'Ivoire	Senegal *	
Djibouti *	Seychelles *	
Equatorial Guinea	Sierra Leone	
Ethiopia *	Somalia *	
Gabon	Sudan*	
The Gambia *	Tanzania	
Ghana	Togo	
Guinea	Uganda	
Guinea Bissau *	Zaire	
Kenya	Zambia *	
Liberia	Zimbabwe *	

Asia

Australia	Thailand
Bangladesh	Vietnam
Bhutan *	
Brunei	* Countries not mapped in this atlas either because insufficient data were available or the areas of rain forest were too small to justify inclusion.
Cambodia	
China	
India	
Indonesia	
Laos	+ Including Cuba, Jamaica, Haiti, Dominican Republic, the Windward Islands, Trinidad & Tobago
Malaysia	
Myanma (Burma)	
Nepal*	
Pacific islands *•	• Various Pacific islands have small rain forests, and the Soloman Islands and Fiji are mainly forest covered.
Papua New Guinea	
The Philippines	
Singapore	
Sri Lanka	
Taiwan *	

0°   60°   120°

97

The making of the maps

The data for the maps comes from three different sources: on-the-ground fieldwork, aerial photographs, and satellite images. Fieldwork is the traditional way to monitor forests, and it is still vital to establish the accuracy of the more remote data obtained from aircraft or satellites.

Aeroplanes equipped with survey cameras have been the basis of much of the mapping of the earth's surface. The camera takes overlapping photographs, and when viewed in pairs through a stereoscope, the terrain is seen in three dimensions and precise height and distance measurements can be made. Today this type of monitoring is being replaced by satellite monitoring which can cover far greater areas in a much shorter period of time.

In addition to cameras, both aeroplanes and satellites carry radar, which builds up images from signals reflected back from the earth's surface. The system has the advantage of being usable regardless of the weather conditions, and can be used both by day and night.

Data from space

Observing the earth's weather was one of the first applications of space satellites. The earliest weather satellite was launched by the United States' National Oceanic and Atmospheric Administration (NOAA) in 1960. Since then a series of earth resources satellites, which produce data of higher resolution than the earlier meteorological satellites, have been launched. The first of these, Landsat 1, was put into orbit in 1972.

The main instrument carried by all the Landsat satellites is the Multispectral Scanner System (MSS), which records the radiation reflected back by features on the ground such as water, roads, buildings and vegetation. Each of these features reflects slightly different wavelengths and can therefore be distinguished on the satellite images. On more recent Landsat satellites, a new instrument was added, the Thematic Mapper (TM), which has greatly improved ground resolution.

The French satellite, Spot, launched in 1985, can identify features as small as 10 metres (33 feet) in size. In addition, the satellite is the first remote-sensing device to possess steerable sensors that can be moved laterally until they detect a gap in the clouds through which to take a reading. This means that they can also collect views of the same patch on the ground from several different points in space. By combining these pictures a composite image can be produced from which it is possible to measure both height and distances.

The data collected by Landsat and Spot satellites are relayed to ground stations in the form of electronic signals which are then processed into false-colour photographic images similar to those of a very high-quality television screen. By means of meteorological satellites that record images twice a day, and Landsat data of better quality available from any given area every 18 days, it should be possible to record changes in the world's forest cover. However, problems with cloud cover, with receiving stations that are not continuously in operation, and with the overall cost using the equipment means that many of the tropical forest areas have been monitored only sparingly and spasmodically.

The maps in this book have been produced using information from all the sources of data mentioned above, from all over the world; it has come in a wide variety of scales and projections, and it dates from different times. In general, virtually all the information comes from published or unpublished reports from the past decade (see page 200). But up-to-date information on parts of Central Africa, including Congo, Equatorial Guinea, Gabon and western Zaire is simply not available. The maps for these regions were produced from a number of more generalized sources.

In a number of cases, national forestry departments have provided hand-coloured blueprints of forest cover, often at 1:1,000,000 or 1:500,000 scale: those of Brunei, Laos and Malaysia are notable examples.

1. The first stage in producing an up-to-date map is to locate on an existing outline map basic information such as relief, rivers and roads. The contours and outlines are "read" off the outline map with the use of an electronic pen or digitizer, which transmits precise coordinates electronically to a computer.

2. The computer logs the coordinates while showing on screen the outline of the area being mapped. No matter what the scale or projection of the outline map, the computer can assimilate all the information and display in a uniform way. Several different types of information – forest type, rainfall, humidity, level of soil nutrients, and so forth – may be added successively from different sources to appear all together on the completed map.

3. When all the information required has been transferred on to the computer a printed copy of the map is produced. It is at this stage that the cartographer can for the first time get an impression of how the final map will look. Adjustments to the colour-coding and types of symbols used can be made. It is also possible to print the map using different projections – the manner in which the rounded surface of the earth is shown in two dimensions – and scales to see which will be the most useful.

4. Careful checking of the finished map against all the sources from which the map was compiled is required to ensure that the information is both correct and readily comprehensible – that the information can be interpreted not just by scientists and geographers but by politicians, policy makers, and other concerned individuals.

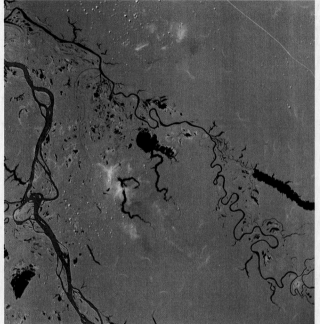

Amazon from space – At the confluence of the Rio Negro, a blackwater river, and the brown, silt-laden Amazon lies the city of Manaus (*left*). Deforestation around the city is visible along with a scattering of clouds. In the false-colour image below rain forest is shown in red, water in blue and silt in black.

Problems with Monitoring

Many countries have invested heavily in remote sensing in the hope that such technology can solve the problems associated with natural resource management and environmental protection. Although considerable progress has been made in the development of the tools themselves – by way of satellites, sensors, data-processing systems and so on – much remains to be done to interpret the information received in a way that is truly relevant to the issues.

The low frequency of observation from Landsat satellites (once every 18 days) in combination with the only too common coincidence of cloud, and the smoke from forest fires, often results in an inability to collect really useful data more than once in every two to four years. In addition, Landsat may obtain the data in a seemingly quite unsystematic manner either because of the unhelpful trajectories of successive orbits or because, for instance, there may be too few operational receiving stations within range. Massive forest fires that raged through tropical forest in Borneo in 1982 and 1983 went essentially undetected and unreported until they were well advanced. Moreover, although data from the Advanced Very-High-Resolution Radiometer (AVHRR) were routinely recorded in the Jakarta-Pakayon station at that time, they were not analysed – they were not even retained for more than a few days – so neither scientists nor officials were alerted to the true extent of the conflagration.

Finally, because of the relatively high resolution and limited ground coverage (more than 200 Landsat scans would be necessary to cover the Amazon Basin), an exhaustive survey using this platform is inevitably difficult and expensive to carry out. In many instances, the people who really need the satellite images – research scientists, Third World governments and environmental agencies, for example – simply cannot afford the pictures from space.

In comparison, the data being collected daily from the AVHRR carried on the NOAA's polar orbiting meteorological satellites generally include a series of images which provide at least one cloud-free image of most tropical areas during each year's dry season. However, the one-kilometre ground resolution of AVHRR is suitable for identifying only large-scale deforestation.

There have been a number of attempts to identify disturbances within the tropical rain forest using satellite images. Deforestation in the southern fringe of the Amazon Basin, in the state of Rondônia, has been assessed by reviewing a series of time-lapse AVHRR data collected since 1982. The measurement of deforested areas there depended upon the preliminary detection of "disturbance areas" in which activities such as road building, mining, logging, clearing for agriculture and burning were taking place. By using thermal sensing systems, areas of higher temperatures could be seen on the satellite images, created by the removal of the forest canopy and the resultant greater absorption of the sun's heat. The data has provided useful information on the rate of deforestation there.

To date, then, satellite remote sensing cannot completely replace conventional mapping and forest classification using aerial photography and fieldwork on the ground. The assessment of forest degradation likewise still requires careful groundwork. Nevertheless, remote sensing could be used much more effectively than is the case at present. One particular problem is that governmental and international remote-sensing agencies around the world tend to study rain forests piecemeal. In view of the significance of tropical deforestation to all people on earth, the time is right for the resources to be made available for an international and comprehensive initiative to map and monitor the rain forests of the world.

Information about the Caribbean was particularly difficult to obtain, especially in view of the considerable uncertainty as to the original extent of the rain forests. Useful sources for information on the Lesser Antilles comprised a series of data atlases prepared by the Eastern Caribbean Natural Area Management Programme. For various other parts of the Caribbean some admittedly rather dated maps were used as a basis for assessing the dimensions of recent alterations in forest cover from local correspondents.

Mapping different types of forest

The maps in this book show four types of rain forest: lowland, mangrove, montane and former rain forest. These categories are the result of compilation, simplification and harmonization of a wide range of different maps. National maps of vegetation are invariably much more complex than those presented here, showing a wide range of different types of rain forests. Simplification has its dangers and this section describes as clearly as possible the limitations of the maps on the following pages.

No attempt has been made to distinguish relatively pristine from more or less degraded rain forests. As some of the richest environments on earth, and being relatively easy of access, lowland rain forests are subject to a wide range of impacts, described on pages 36–45. Even a heavily logged rain forest, however, can quickly develop a closed canopy and it is difficult for anyone but an expert to distinguish logged from unlogged forest after a few decades. Similarly, while large areas of plantations and shifting cultivation have been excluded, the scale of mapping does not allow for presentation of all enclaves of agriculture and plantations within the main forest blocks. The areas mapped as lowland rain forests are therefore mosaics of relatively undisturbed forests mixed with often extensive disturbed forests and some, relatively small, patches of cultivation. This applies

equally well to montane rain forests, but in general these have suffered less degradation. The reason is that most montane rain forests do not have such valuable timber trees and are less accessible to both loggers and agriculturalists.

The boundary between lowland and montane forests is drawn at different altitudes on different maps to make some allowance for the so-called "Massenerhebung effect". First recorded in the European Alps, this is the phenomenon whereby large mountains and the central parts of large ranges are warmer at a given altitude than small mountains and outlying spurs. The consequence of this is that on small or more exposed mountains montane forest (described on page 24–25) is found lower down. This is why the maps for Madagascar, Australia and Southeast Asia show montane forest beginning at altitudes of 910 metres (3,000 feet), whereas in West Africa the cut-off point is 1,200 metres (4,000 feet), in New Guinea it is 1,400 metres (4,500 feet) and in South America it is 1,800 metres (6,000 feet).

Semi-evergreen rain forests grade into seasonal or monsoon forests where the climate is highly seasonal, and the majority of trees lose their leaves in the dry season. These seasonal forests have generally not been included in the maps. They are present on all continents and quite extensive in Asia, but they are not easily delimited from the thorn forests and more open woodlands at their drier limits, and in places they occur as a mosaic with rain forests. Where they are naturally found, they have been mapped as outside the rain forest, but where they result from degradation they are included in former rain forest.

Assessing the areas of former rain forest contains its own difficulties. Botanists have studied climatic data, forest outliers and soil types to reach reliable conclusions, but many uncertainties remain. In the maps of Africa there is a category "Mosaic of grassland and former rain forest". In this area it is impossible to know precisely where rain forests once grew.

■ Lowland Rain Forest

■ Mangrove Forest

■ Montane Rain Forest

□ Former Rain Forest

Lowland rain forests

These forests are evergreen or semi-evergreen, closed-canopy rain forests that occur in ever-wet climates. Here the rainfall is evenly distributed and the dry season, if it exists at all, is very short and does not cause the trees to lose their leaves. This atlas deals with *tropical* rain forests, but in northern Myanma (Burma), the foothills of the Himalaya and the Andes there is a gradation of rain forests from tropical to temperate, and there is necessarily a degree of arbitrariness in the cut-off point.

Mangrove forest

Mangrove is seldom extensive, but is common as a narrow coastal and estuarine fringe. It is often well-mapped, being relatively easy to identify from aerial photographs. Mangroves are particularly important to rural communities in providing a number of commodities (see pages 22–23). Worldwide they are being cleared to make way for prawn or fish ponds and housing. More recently they are being clear-felled for paper pulp and rayon manufacture.

Montane rain forests

This forest category includes lower montane and upper montane forests, as well as the subalpine formations on the highest mountains. In many parts of the world montane rain forests remain intact, but this is by no means always so. The montane rain forests of East Africa are rich in valuable timbers and have been heavily logged. In Papua New Guinea montane forests have attracted shifting cultivators because sweet potatoes, a staple crop, grow well there.

Former rain forest

The precise extent of former rain forests is clearly a matter of conjecture. In some areas – Java is an example – they were cleared thousands of years ago. Former lowland and montane rain forests as well as former mangrove forests are included in this category. Where rain forests are believed to have degraded into seasonal monsoon forests, as in mainland Southeast Asia, they are also included in this category.

State boundaries – These are shown as a continuous evenly-dotted line.

Physical features – The names of major mountain ranges, hills, rivers and other physical features are given where appropriate. The names of some major cities are also included.

Not rain forest – The grey/white areas on the map denote a wide variety of natural and man-made habitats that are outside the rain forest or former rain forest zones. They include areas of different types of forest, such as deciduous seasonal forests and plantation forests, as well as woodlands and grasslands – and pockets of any of these may account for apparently anomalous gaps in coloration on the map.

Roads – Only certain major trunk routes are shown.

Relief – Some indication of uplands and lowlands is given by the differing shades of grey: the darker the grey, the steeper the terrain.

Protected areas – There are more than 5,400 protected areas in the world, almost 2,000 of them in tropical countries. To map them all would require a gazetteer devoted entirely to this subject. Accordingly, only those protected areas mentioned in the text are shown (red hatching) and named. Areas that are likely to be placed under official protection in the future, and that are mentioned in the text, are indicated by dotted hatching.

National boundary – Alternate short- and long-dotted lines are used to show national boundaries.

Map within a map – Each map features an additional locator map showing the region of the world in which it is located.

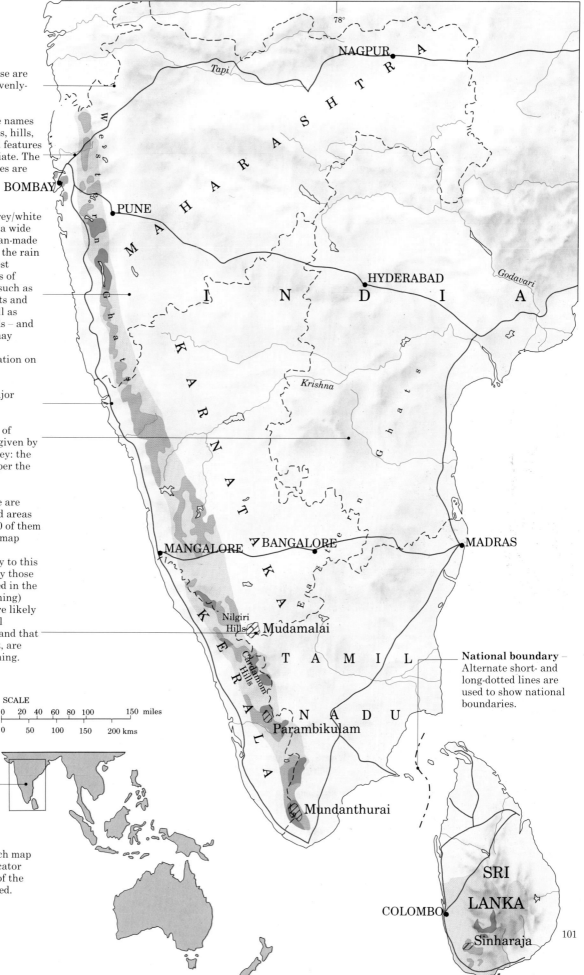

SCALE

0 20 40 60 80 100   150 miles

0   50   100   150   200 kms

Central America

The seven Central American countries south of Mexico – Belize, Guatemala, Honduras, El Salvador, Nicaragua, Costa Rica and Panama – although each small in size, contain a unique concentration of flora, fauna and people. As a result of their location on the land bridge between the vast and very different continental ecosystems of North and South America, the relatively tiny tropical forests in this region are among the richest habitats on earth in terms of the number of species they contain. Even the smallest of the seven countries, El Salvador, has more than 350 bird species. Panama tops the list with 700 – more than the whole of North America. The rain forests in southern Mexico represent the northernmost extent of this habitat.

The human population, which has doubled over the last 30 years, is a mix of many different ethnic groups and, combined with an unstable political situation, is placing ever-increasing pressure on the rain forests. Before 1950, the lowland and submontane forests were virtually intact. This changed very rapidly. In 1950 about 60 percent of Central America was covered with forest or woodland. By 1970 this was estimated to have fallen to 49 percent, and a mere decade later it had shrunk still further to 41 percent. At the current rate of deforestation, most of the remaining forest will be eradicated within the next 20 years, leaving only impoverished remnants in reserves and national parks. Already El Salvador has virtually no primary forest left and, with the exception of Belize where development pressure remains comparatively low, the other countries in the area are losing between 500 and 1,000 square kilometres (190 and 390 square miles) of forest a year.

In percentage terms, the annual rate of loss in much of Central America far exceeds that of countries such as Brazil and Malaysia.

Although the forest losses in Central America are small on a global scale, they represent an astounding rate of change for the countries concerned. Indeed, in percentage terms, the annual rate of loss in much of Central America far exceeds that of countries such as Brazil and Malaysia, which are generally cited as undergoing rapid deforestation.

Population pressure

The combined population of Central America (excluding Mexico) in 1989 was almost 30 million people, more than double that of 1960, and it is expected to increase to about 70 million by the year 2025. Most people throughout the region are of mixed European, indigenous Indian, African and West Indian blood; however, all the countries (except Belize) are dominated by a Spanish-speaking Westernized elite, and large unintegrated indigenous populations still exist in many areas. The distribution of the population is very uneven. The majority of the people inhabit the volcanic montane regions and intermontane valleys, more than two-thirds of them living within 65 kilometres (40 miles) or so of the Pacific Ocean.

El Salvador is one of the most densely populated countries in the world, whereas virtually all of Belize and the eastern lowland areas of Honduras and Nicaragua are among the most sparsely inhabited areas in the Western hemisphere. Many of the people live in poverty and have either no land or not enough to support a family. In contrast, a minority of extremely wealthy landowners controls most of the productive areas.

The names of many of the Central American countries are synonymous with political and social unrest, poverty and war. In spite of this instability, many ambitious improvement pro-

Mexico – Within the past 35 years more than half of Mexico's rain forest has been lost. Today, about 150,000 sq km (58,000 sq miles) of forest remains. Mexico's population of 87 million people is increasing at a rate of about 2 million people a year. In response to this, much of the forest has been cleared as part of government-sponsored resettlement programmes, cattle ranching projects and coffee plantations. There is also much uncontrolled clearance as settlers clear land for themselves.

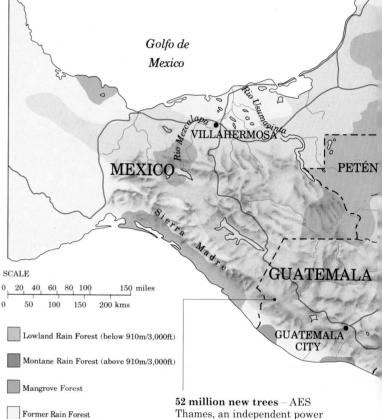

SCALE

0 20 40 60 80 100   150 miles

0   50   100   150   200 kms

- Lowland Rain Forest (below 910m/3,000ft)
- Montane Rain Forest (above 910m/3,000ft)
- Mangrove Forest
- Former Rain Forest
- Protected Area (referred to in text)

52 million new trees – AES Thames, an independent power producer in the United States, is spending US$2 million on planting 52 million trees in Guatemala to offset the effects of carbon dioxide, a "greenhouse gas", released from one of its coal-fired power stations. Through a grant to CARE (Cooperative for American Relief Everywhere) Inc., the money will be used to help 40,000 smallholders to plant the trees over a 10-year period. This is intended not only to help reduce global warming, but also to provide a sustainable yield of fuel, food and building materials for the local people and to reduce soil erosion.

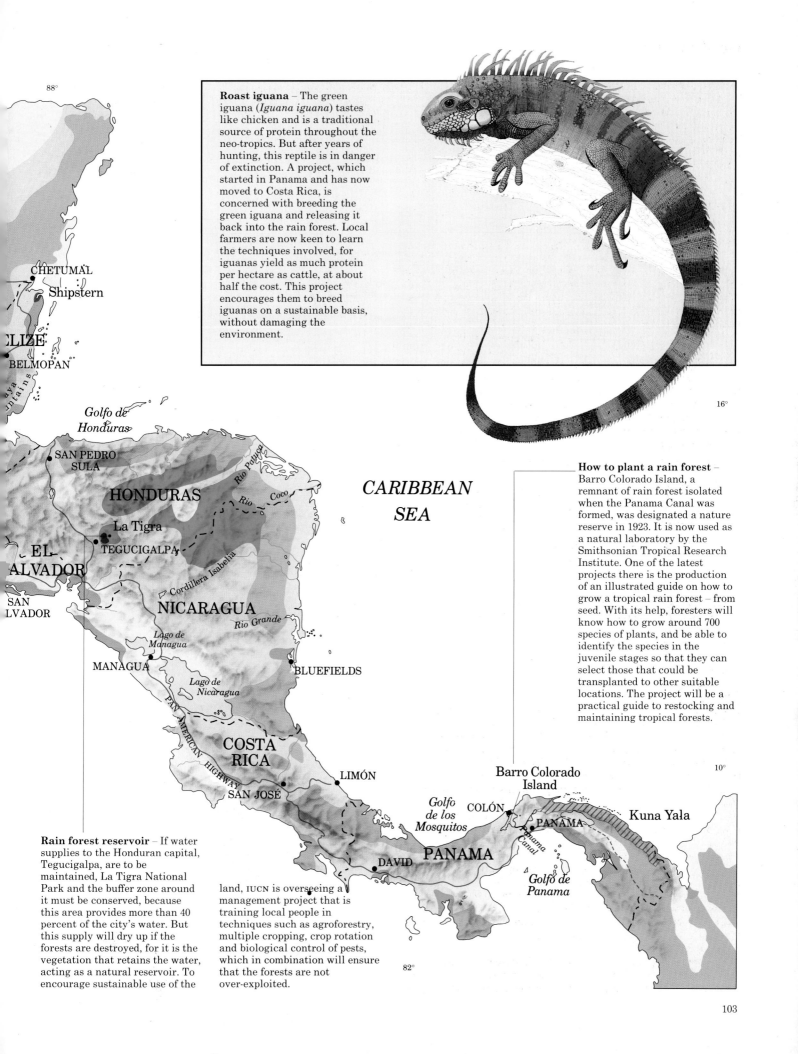

Roast iguana – The green iguana (*Iguana iguana*) tastes like chicken and is a traditional source of protein throughout the neo-tropics. But after years of hunting, this reptile is in danger of extinction. A project, which started in Panama and has now moved to Costa Rica, is concerned with breeding the green iguana and releasing it back into the rain forest. Local farmers are now keen to learn the techniques involved, for iguanas yield as much protein per hectare as cattle, at about half the cost. This project encourages them to breed iguanas on a sustainable basis, without damaging the environment.

88°

CHETUMAL
Shipstern

ELIZE
BELMOPAN

aya
untains

Golfo de
Honduras

SAN PEDRO
SULA

HONDURAS

Río Patuca

Río   Coco

La Tigra

EL
ALVADOR

TEGUCIGALPA

Cordillera Isabelia

SAN
LVADOR

NICARAGUA

Río Grande

Lago de
Managua

MANAGUA

BLUEFIELDS

Lago de
Nicaragua

PAN AMERICAN HIGHWAY

COSTA
RICA

LIMÓN

SAN JOSÉ

CARIBBEAN
SEA

16°

How to plant a rain forest – Barro Colorado Island, a remnant of rain forest isolated when the Panama Canal was formed, was designated a nature reserve in 1923. It is now used as a natural laboratory by the Smithsonian Tropical Research Institute. One of the latest projects there is the production of an illustrated guide on how to grow a tropical rain forest – from seed. With its help, foresters will know how to grow around 700 species of plants, and be able to identify the species in the juvenile stages so that they can select those that could be transplanted to other suitable locations. The project will be a practical guide to restocking and maintaining tropical forests.

10°

Barro Colorado
Island

Golfo
de los
Mosquitos

COLÓN

PANAMA

Kuna Yala

Panama
Canal

DAVID

PANAMA

Golfo de
Panama

Rain forest reservoir – If water supplies to the Honduran capital, Tegucigalpa, are to be maintained, La Tigra National Park and the buffer zone around it must be conserved, because this area provides more than 40 percent of the city's water. But this supply will dry up if the forests are destroyed, for it is the vegetation that retains the water, acting as a natural reservoir. To encourage sustainable use of the land, IUCN is overseeing a management project that is training local people in techniques such as agroforestry, multiple cropping, crop rotation and biological control of pests, which in combination will ensure that the forests are not over-exploited.

82°

grammes are under way to increase agricultural productivity, increase exports, stimulate industrial growth or provide roads to remote areas. Many of these have been undertaken at the expense of the forests and other natural resources in the area. The projects may solve immediate requirements for extra food, more employment and increased revenue, but the consequences of the over-exploitation of the forests include soil erosion, sedimentation of dams and harbours, and water pollution.

Squandering the timber

The landless settlers are responsible for much of the forest destruction as they try to eke out a living from the soil. Initially, logging or government development companies build roads into unpopulated areas; this is followed by the arrival of peasant farmers who cut down the trees and plant crops in their place. For example, in Guatemala the Government built a road into the Petén region in the north of the country, and then encouraged the farmers that moved there to grow coffee, cardamom, cacao and rubber. The timber that is cut to make way for these crops is generally burned or left to rot instead of being harvested, compounding the waste of potentially valuable resources. In Honduras alone, it has been estimated that forests with a commercial timber value of US$320 million are squandered

PACIFIC OCEAN

ATLANTIC OCEAN

→ Main bird migration routes from North to Central and South America

Butterflies for cash – In the Shipstern Reserve in Belize butterflies are being used to provide funds for conservation. This reserve is situated in one of Central America's most important wetlands, and contains tropical moist forest, savanna and lagoons on the coast near the border with Mexico. The main objective of the Reserve is to show that the sustainable use of forest resources can provide enough income to support the area. Initially, marketing of captive-bred butterfly pupae, including the zebra longwing (*Heliconius charitonius*), malachite (*Siproete stelenes*) and Thoas swallowtail (*Papilio thoas*) pictured above, will provide the money for further work.

annually in this way. Much of the forest soil is shallow and of limited fertility and this, combined with invasive weeds and noxious insects, soon drives the peasants on to clear yet more land elsewhere.

The degraded land is then sold or taken over by speculators and cattle ranchers who consolidate the smallholdings into larger ones for the exclusive purpose of raising beef cattle. For a few years, each hectare (2.5 acres) will support one head of cattle but, within five to ten years, this has increased to five to seven hectares (12 to 17 acres) per animal. Ultimately the land will be left to scrub and secondary growth or, even worse, the infertile, denuded soil will be lost through erosion.

The farming methods used by the rain forest Indians are invariably more productive than the pasturalism that is replacing them. In addition, neglect of the old cultural techniques, such as the construction of terraces and contour planting on steep slopes, is contributing to the widespread erosion and soil degradation throughout Central America.

The disappearing migrants – The destruction of the rain forests in Central America may well be having a more widespread effect than was initially thought. The region is used by at least 225 different species of migratory birds from both North and South America. In fact, three of the four migration routes between the two Americas converge on Panama, flyways for land birds, seabirds, waders and waterfowl all passing over or near the area. A number of scientific studies have suggested that there is a connection between the widespread deforestation in Central America and the decline of certain of the common North American migrant species. About one-third of the 53 bird species that winter in Central America have been decreasing in number; and many of these are the ones that spend their nonbreeding season on the Pacific coast of Central America. It is this area that has suffered the greatest forest loss in recent years. Well documented examples of the decline in bird numbers include those of the wood thrush (*Hylocichla mustelina*), the chestnut-sided warbler (*Dendroica pensylvanica*), the Tennessee warbler (*Vermivora peregrina*) and the black-throated green warbler (*Dendroica virens*). In addition to deforestation, the profligate use of pesticides in the highland and Pacific coastal areas of Central America may also be playing a part. A more traditional threat to the birds comes from the Quiche Indians in western Guatemala, who light huge bonfires at night to attract, trap and kill them for food. But this has been going on for a long time, and it is more likely to be the reduction of the birds' wintering habitats through deforestation that is the major factor in causing their decline.

Continuous dredging is necessary to keep the canal open.

The Panama Canal: no forest, no canal

Completed in 1914, the Panama Canal is a masterpiece of engineering technology and has for years been one of the most important shipping lanes in the world. However, its very existence is being threatened by destruction of the rain forest on the surrounding hillsides. About 12,000 ships use the canal annually, and as each passes through the series of locks on the waterway, almost 200 million litres (53 million US gallons) of water flow out to sea. This water is supplied by the heavy rains that fall on the mountain ranges surrounding the canal. Although the rain does not fall all year round, this does not affect the canal as the forest within its watershed acts as a giant sponge, soaking up the rainwater and releasing it at a steady rate throughout the year.

In the late 1970s there was, for the first time, a drastic drop in water level in the canal. The government of Panama had no option but to turn away the largest ships, thereby losing income from the canal tolls, which amount to US $350 million a year. The reason for this was, without doubt, the loss of the forest; settlers are cutting down between 30 and 50 square kilometres (11.5 and 19 square miles) of forest within the watershed each year. To compound the problem, with no tree roots to hold the soil in place, the heavy rains wash it away, only to silt up the canal. A strict law against logging, forbidding the felling of trees more than five years old, was introduced in Panama in 1987, but the government has no money to enforce its own legislation. So when Panama takes full control of the canal in the year 2000 it is likely to inherit a waterway open only to the smallest of ships.

Without trees to protect the soil landslides are common.

Kuna: organizing to survive

The Kuna Indians of Panama are the world's first indigenous people to have established an internationally recognized forest park, to protect their culture and land. As a result, the Kuna Yala Reserve has become a *cause célèbre* among environmentalists and advocates of indigenous peoples' rights.

About 30,000 Kuna Indians live in 60 villages on some of the 350 tiny coral islands off the east coast of Panama. They survive by a combination of fishing and subsistence farming along the coast, and so have not touched the main part of the forest on the mainland. Indeed, farming, hunting and the felling of trees are specifically prohibited in a number of spirit sites near the villages. The Kuna, unlike many indigenous populations, own the land they live on and have done so since the 1930s, when they forced the government through a series of armed uprisings to set aside the land for them.

In the 1970s, the construction of a new road at the western end of the Kuna reservation began. Although this was initially welcomed by the Indians, it became evident that squatters would soon move in and take over large tracts of the Indians' land. To prevent this happening, advice and financial backing was sought by the politically astute Kuna from a number of national and international organizations. As a result, the Kuna Wildlands Project, known as PEMASKY, was set up and it may eventually be designated a biosphere reserve.

Kuna in control

The design, implementation and management of the project is firmly in the hands of the Indians. They receive the revenue from tourists and the scientists, who have to pay to work or visit the area. In addition, each scientist has to be accompanied by a Kuna assistant and has to leave a record of his findings with the Indians. The project provides funds for the Indians, they benefit from the research done there and, perhaps most importantly, the forests and the heritage of the Kuna Indians are being conserved.

The success of PEMASKY has already made it a model for other indigenous groups in Central America. The Embera, who live in the adjacent Darién region of Panama, have begun consulting with the Kuna to develop a similar plan for their lands. In this way, more tropical forests will be preserved to act as an essential and sustainable source of food, medicine, fuel and building materials for its indigenous human occupants, as well as to provide a home for thousands of plants and animals.

> **The Kuna Yala Reserve has become a *cause célèbre* among environmentalists and advocates of indigenous peoples' rights.**

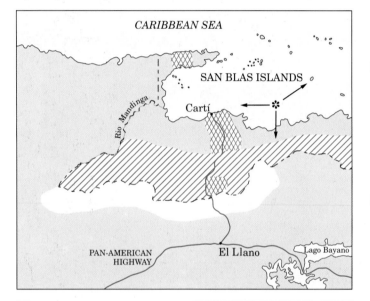

Core Area

Agricultural Area

✳ Cultural Area

Buffer Zone

— Road

PEMASKY – The Kuna Wildlands Project, or PEMASKY, which is centred around the road from El Llano to Cartí, is designed to protect the western end of the Kuna Yala Reserve from settlers, who use the road to gain access to otherwise inaccessible Kuna land. The Project contains four categories of protected area: a core area in which only tourism and scientific research are permitted; an area reserved for Kuna agriculture; a cultural area, which includes the islands on which the Kuna live, their fisheries and coastal agricultural plots; and a restoration area (situated just outside the Reserve borders) which acts as a buffer zone, protecting the Project from outside interference.

Dug-out transport (*above*) – Each morning the Kuna use their dug-out canoes to cross the sea from the coral islands on which they live to the mainland, where they carry on a multi-crop agricultural system on the coast.

Molas (*right*) – Colourful, hand-sewn, appliquéd molas are a common form of traditional dress among Kuna women. The dyes come mainly from natural sources and each mola usually takes about 2 months to make.

The Caribbean

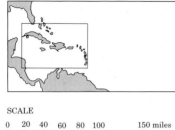

Lowland Rain Forest (below 910m/3,000ft)

Montane Rain Forest (above 910m/3,000ft)

Mangrove Forest

Former Rain Forest

Protected Area (referred to in text)

SCALE

| 0 | 20 | 40 | 60 | 80 | 100 | | 150 miles |

| 0 | 50 | 100 | 150 | 200 kms |

LA HABANA

SANTA CLARA

No data available

Greater

CUBA

HOLGUIN

Sierra Maestra

JAMAICA

Blue Mountains

KINGSTON

Antilles

Cuba – In 1812, 90% of Cuba was covered with forest; by 1959, this had fallen to a mere 14%. The situation remained constant throughout the 1980s because most of the remnants are in the mountains. Rain forest probably once covered the south of the main island, but now only 2 patches of significance remain on the slopes and peaks of Sierra Maestra and Sierra de Imías. Fragments of montane forest, too small to show on the map, still persist in the Sierra del Escambray.

The islands of the Caribbean, stretching in an arc from Florida to Venezuela, are the emergent tops of a chain of ancient volcanoes, some of which are still active. The island geography of the Caribbean has favoured the evolution of many endemic species: Jamaica alone contains four mammal, 26 bird, 27 reptile, 20 amphibian and at least 900 plant species that are found nowhere else in the world. But in common with other oceanic islands, the Caribbean islands have special conservation problems. They lack the natural buffers that are found in large biological communities and hence are more at risk. Island ecosystems, which have evolved in relative isolation, are easily upset.

Before the sixteenth century many of the islands were almost totally covered in forest but now, due to the limited land available and the high population density, most of the forest has disappeared. Many islands, including Barbados, are now planted in sugar cane; others, Dominica in particular, have retained substantial areas of forest. Human impact on the forest has been particularly heavy in the Greater Antilles, but all the islands are losing their forest as it is cleared for agricultural use.

The Caribbean is prone to hurricanes, earthquakes and volcanic activity, which destroy large tracts of forest. In 1979, hurricanes David and Frederick killed more than 2,000 people in the Dominican Republic and caused damage worth nearly US$1,000 million. Severe secondary effects of hurricanes include damage to agriculture, which can lead to famine and economic disaster for small islands dependent on one or two export crops. The hurricane damage is worse in deforested areas because the exposed land is susceptible to landslides and floods.

There is one island where the rain forests are not uprooted by hurricanes; instead they are moulded by them. This occurs in the mountainous region of Guadeloupe, in the Basse Terre National Park. The forest canopy here is merely trimmed by the winds so that, from a distance, it resembles a well-cut lawn.

The Caribbean Conservation Association, the leading non-governmental organization in the area, is concerned with all aspects of protecting the environment. Backed by member governments and international funding bodies, the Association has been involved in identifying key watersheds, potential parks, protected areas and important habitats for endangered species. It is involved, among other things, with environmental education, pilot efforts in management of timber and marine resources, and in developing appropriate legislation. Income from sustainable development of the land and sea, wildlife tourism and harvesting of forest products will enable the Caribbean islands to prosper without destroying more of their rain forests.

Dominica – This island is the most mountainous and the most forested of any in the Lesser Antilles. The 80,000 inhabitants, mostly descendants of slaves brought from Africa for plantation labour, are concentrated along the coast, leaving nearly three-quarters of the island undisturbed and covered in forest. However, clearance of the forest for agricultural land has increased in the last 5–10 years, particularly as many local farmers now have access to chainsaws. In addition, in 1979 Hurricane David caused the most severe damage ever reported to forests on the island. No part of Dominica escaped and it was estimated that up to 5 million trees were damaged. Hurricanes not only uproot trees and tear off branches, they also disrupt fruiting and flowering cycles, which, in turn, affects the animals in the area. The rain forests will probably take more than 50 years to recover fully.

Puerto Rica – This island provides a case study for the natural rehabilitation of a severely damaged environment. In 1508, when Europeans first arrived, Puerto Rica was almost completely forested. By 1770, only 6% of the island's 8,630 sq km (3,330 sq miles) had been cleared; but after 1815, when the island was first opened to international trade, clearing began in earnest and did not stop until 1903, by which time a mere 0.4% of the land was still forested. With the collapse of the sugar cane industry in the 1940s, much of the land was abandoned. There followed a dramatic natural increase in secondary forest, so that by 1978 it covered 32% of the island.

A source of pride (*left*) – The St Lucia parrot (*Amazona versicolor*) is found only in remnant patches of forest in the interior of the island, and is on the verge of extinction. Initially widespread on St Lucia, the population was down to 1,000 birds by 1950, and further declined to 100 by 1977. They were hunted for food and the pet trade, and their forest habitat was destroyed by human activity and hurricanes. In 1979, the parrot was declared the national bird of St Lucia and has since become an object of pride. Along with the abolition of hunting, the establishment of a forest reserve and a captive breeding programme, this change in attitude may ensure the survival of this handsome bird.

Natural destruction (*left*) – Hurricanes are a major cause of destruction in the Caribbean, and the forests are often badly damaged. During a storm, forested land is less at risk from landslides and flooding because the trees protect the soil from the full impact of the rain.

Trinidad and Tobago – A recent study sponsored by the International Tropical Timber Organization (ITTO) showed that almost none of the countries with tropical forests managed them on a sustainable basis. Trinidad and Tobago are exceptions to this. Trinidad has a long history of forest protection, management and tree cultivation. As long as half a century ago foresters there produced a variety of teak that was well acclimatized to survive outside its native Myanma (Burma) and Thailand. The forest policy of the government of Trinidad and Tobago underscores the need for permanent forest reserves to preserve water supplies, to prevent soil erosion and flooding, and to produce timber and many other valuable forest products.

SANTIAGO

DOMINICAN REPUBLIC

SANTO DOMINGO

AITI

RT-AU-NCE

ata able

Leeward Islands

SAN JUAN

Virgin Islands

PUERTO RICO

Lesser

ST.KITTS-NEVIS

ANTIGUA AND BARBUDA

MONTSERRAT

GUADELOUPE
Basse Terre

Windward Islands

Antilles

DOMINICA

MARTINIQUE

ST.LUCIA

ST.VINCENT

BARBADOS

GRENADA

TRINIDAD AND TOBAGO

PORT OF SPAIN

The Amazon Basin

The Amazon Basin contains by far the largest area of tropical forest in the world, covering six million square kilometres (2.3 million square miles) in nine different countries – 60 percent in Brazil, and the rest in French Guiana, Surinam, Guyana, Venezuela, Colombia, Ecuador, Peru and Bolivia. Biologically, it is probably the richest and most diverse region in the world, containing about 20 percent of all higher plant species, perhaps the same proportion of bird species and around ten percent of the world's mammals. Each type of tree may support more than 400 insect species. Much of the Amazon Basin remains unknown and each expedition there seems to discover something new.

At the heart of this region lies the huge Amazon River, second longest in the world, which has at least 1,000 tributaries and holds more than one-fifth of the earth's fresh water. It discharges as much water into the sea each day as London's Thames does in a year. The Amazon is estimated to contain 2,000 species of fish, ten times the number in European rivers, not to mention numerous reptiles and mammals including the Amazonian manatee (*Trichechus inunguis*), the spectacled caiman (*Caiman crocodilus*), the giant anaconda (*Eunectes murinus*) and the Pink River dolphin (*Inia geoffrensis*) or bôto. In flood, it inundates an area the size of England creating a unique habitat, the flooded forest. Higher up the Amazon watershed, where the the Amazon's tributaries flood for a short time after heavy rains, the flooded forest is called *várzea*; farther downstream, in the Amazonian lowlands, there is the swamp forest, or *igapó*, which is usually flooded for between four and seven months each year.

The burnings

The scale of forest destruction has become so great in the seasonal and rain forests of Brazil that Brazilians now talk of three seasons: the rainy season, the dry season and the *queimadas*, or burnings. Neither the actual extent nor the rate of deforestation in any of the Amazonian countries is known for sure. Many figures have been produced but estimates vary

> **. . . Brazilians now talk of three seasons: the rainy season, the dry season and the *queimadas*, or burnings.**

hugely. Most widely quoted are those given by the Food and Agriculture Organization (FAO), which estimated that 25,300 square kilometres (9,800 square miles) of Brazilian rain forest was destroyed each year between 1981 and 1985. However, the latest figure produced by Brazil's National Institute for Space Research was of a loss of 35,000 square kilometres (13,500 square miles) a year in 1987 and 1988. That is equivalent to four square kilometres (1.5 square miles) every hour.

Other recent FAO estimates include a loss of 6,000 square kilometres (2,300 square miles) a year in Colombia, 3,400 square kilometres (1,300 square miles) a year in Ecuador, 2,600 square kilometres (1,000 square miles) a year in Peru, while the other Amazonian countries are each losing hundreds of square kilometres of rain forest a year.

In 1885, Baron de Santa-Anna Néry described Amazonia as "the virgin soil which awaits the seed of civilization". Today, cattle ranchers, industrialists, and even international money-lending organizations sometimes still act as if that is what they think is the best prospect for the region. The causes of the deforestation are many, although of overriding importance is conversion to pasture land for cattle ranching and, secondly,

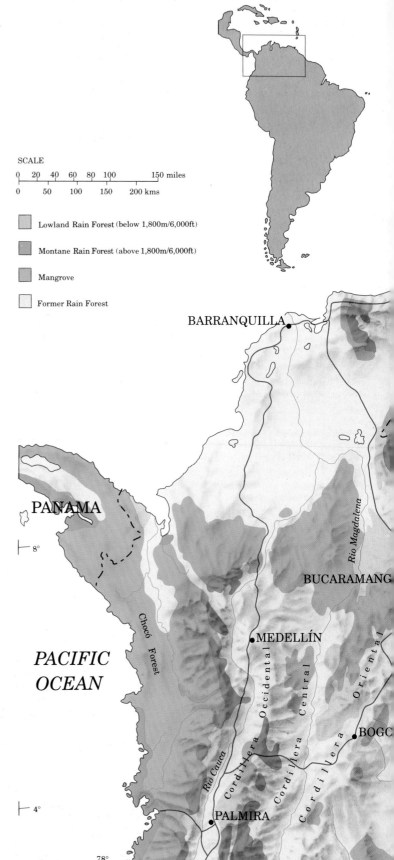

SCALE

| 0 | 20 | 40 | 60 | 80 | 100 | | 150 miles |

| 0 | 50 | 100 | 150 | 200 kms |

Lowland Rain Forest (below 1,800m/6,000ft)

Montane Rain Forest (above 1,800m/6,000ft)

Mangrove

Former Rain Forest

Palmira – For the past 40 years, people have settled on the forested slopes above the city of Palmira. Much of the forest, which is vital for the protection of the city's water and hydroelectric energy supplies, has been converted to pasture, with the result that the land is becoming severely eroded. Now the local power company, is helping to replant the area.

Colombia's Chocó – The stretch of lowland forest along Colombia's coast, known as the Chocó, is estimated to contain 8,000–9,000 plant species, one-quarter of which are found only here. The Chocó also has more than 100 bird species unique to the area. These forests are increasingly threatened by logging, settlement and agricultural development.

Tree of life (*left*) – The Buriti palm (*Mauritia flexuosa*) is a veritable "tree of life" to many Amazonian Indians. Its fruit is as rich in vitamin C as an orange, its pulp oil richer in vitamin A than spinach. The Indians use starch extracted from the pith to make bread, and the fruit, sap and inflorescences to make wine. They also eat the palm heart. In addition, a strong fibre is obtained from the young leaves, a useful cork-like material is extracted from the petioles, and the wood is used for building.

CARIBBEAN SEA

12°

TRINIDAD AND TOBAGO

PORT OF SPAIN

MARACAIBO

VALENCIA

CARACAS

Lago de Maracaibo

Cordillera de Mérida

VENEZUELA

PAN AMERICAN HIGHWAY

8°

Rio Orinoco

CIUDAD BOLIVAR

SAN CRISTÓBAL

Rio Apure

Rio Arauca

Rio Caura

Rio Meta

COLOMBIA

4°

Rio Guaviare

AMAZONAS TERRITORY

BRAZIL

72°

66°

small-scale agricultural settlement. In contrast to the forests of Southeast Asia and West Africa, logging has not been of great significance in the Amazon Basin.

Inappropriate land use, often based on poor knowledge of the ecology and soils of the region, has caused, and is still causing, many problems. Intensive systems of agriculture and exotic crops have been used in areas where the soil is unable to support them on a long-term basis. Future plans need to be based on a much more detailed knowledge of the potential for development, so that the land can be used in appropriate ways. But although it is slowly being realized that the forests on the poorer soils can be harvested sustainably for such commodities as rubber, oils, nuts, palm hearts and timber, the deforestation continues.

Looking further afield

Destruction of the Amazon rain forest will have an effect far beyond the borders of the nine Amazonian countries. Because of the huge volume of cloud it generates, the Amazon system is believed to play a major role in distributing the sun's heat around the globe. In a way that is not yet fully understood, cutting of the rain forest also causes less rain to fall in the Amazon Basin itself (see page 35).

It is obviously unreasonable for the developed countries to expect Brazil and the other Amazonian countries to leave their forests totally untouched, especially since it is the developed nations that are demanding the beef, the hardwood, the rubber, the cocaine and the oil from the region. Most of the Latin American countries have serious economic problems, including vast foreign debts – the combined external debt in 1987 of Brazil, Peru, Venezuela, Ecuador, Colombia and Bolivia was more than US$2 billion. Their poor people are demanding a better way of life and they have little spare money to pay to conserve and study the rain forests. The developed countries can therefore help by finding ways of financing solutions to the problems.

Brazil: gaining access to the forest

The first inhabitants of the Amazonian rain forest, the Amerindians, migrated from the north, across the isthmus of Panama, at least 20,000 years ago and reached the Amazon Basin about 10,000 years later. The size of the Amerindian population at the time of the arrival of Europeans in the sixteenth century is debatable; it may have been as high as 15 million or as low as two million people. Whatever the initial figure, now only 200,000 Indians remain; the rest died as a result of the diseases, enslavement and social disruption brought by the Europeans. More than one-third of the tribes present in 1900 have since disappeared.

The main settlement in the Brazilian Amazon began in the mid-1970s when the then president, Emilio Medici, built the Transamazonia Highway as part of Operation Amazonia. Easily forgotten in the current environment-conscious climate is the fact that this road was built with the world's approval. Operation Amazonia was seen as a way of opening up much needed land for poverty-stricken Brazilians, enabling the development of "a land without men for men without land". That there were "men without land" was not due to huge population growth, nor was it because there was insufficient cleared land available for cultivation, instead it was, and still is, because 43 percent of Brazil's most productive land is owned by one percent of its population.

Once the Transamazonia Highway was built, the landless peasants moved in and cleared as much forest as possible. They were given title only to "productive land", and deforestation was considered proof that they would be growing crops on the land. Once the settlers moved in they generally received no advice on what crops to grow or how to grow them. Monocultures were encouraged in spite of the fact that they are the most liable to attack by disease and pests, and the least likely to be sustainable. This was discovered as early as the 1920s when Henry Ford tried to plant huge areas of rubber trees. Deprived of shade and susceptible to diseases, the trees soon withered and died.

Altamira (*above, right*) – In early 1989, Altamira became the meeting-place for a huge gathering of local Indians, including the Xicrin woman pictured above and the Kayapó (*right*), protesting against plans for dams to be built on the Xingú River. If the project had gone ahead, thousands of square kilometres of rain forest would have been inundated and the lives of many Indians devastated. As a result of this meeting, and the international protest that accompanied it, the Brazilian government agreed to drop its plans and the World Bank withdrew its funding.

Forest fish (*above*) – One of the most curious features of the Amazon is the *igapó*, or flooded swamp forest, which may be under as much as 12 m (40 ft) of water for up to 11 months of the year. Swimming among the trees are a profusion of fish including the piranha (of the family Characidae), pictured above, many of which live on the seeds and fruit falling from the canopy above. Another seed-eating fish is the tambaqui (*Collossoma macropomum*) which probably locates the fruiting trees by smell. It then uses its huge molars, more like horse's teeth than those of a fish, to crush the fallen seeds. The fat reserves that the tambaqui builds up while the trees are in fruit enable it to live during the season of low water, when little food is available. Perhaps one of the strangest fish in the *igapó* is the arowana. Up to 1 m (3.3 ft) long, with a heavily-armed mouth, it looks as if it should feed on other fish. In fact the most important component of its diet is large beetles. Many of these are caught by the arowana's leaping a metre or more out of the water and grabbing the beetle off a low-hanging branch. The male fish cannot feed for a month or so each year when guarding its offspring, which it does by holding them in its mouth. Fishermen, collecting for the aquarium trade, take advantage of this behaviour: on finding a male arowana, they kill it by decapitation, to prevent it from swallowing the youngsters as it does when netted. They then scoop up the young fish as they flee from the dead father. The floodplains of the Amazon, with their vast numbers of fish living off the fruits of the forest, offer one of the most outstanding opportunities to use the tropical forest in a rational, productive and sustainable way. Forest fish could become an important source of protein in the future.

> ## It seems that returning the degraded land to rain forest will be a long and hard task.

Reclaiming the pastures

Understanding the mysteries of forest regeneration is the aim of a research project that has been set up at Paragominas, in the state of Pará in northeast Brazil. It will try to discover why land cleared of forest for pasture often becomes degraded and able to support only shrubs, weeds and straggling grasses – the forest rarely regrows.

There appear to be several constraints on regeneration, lack of seeds in the soil being the major problem. Researchers found that cleared land contained no tree seeds, and that the birds, which act as seed dispersers, do not venture far into the grassland. Another problem was predation on the seeds or young plants. Many of the seeds in an experimental plot were carried off by ants or rodents, and the seedlings were destroyed by leaf-cutter ants. In addition, the tree seeds appeared unable to adapt to the harsher growing conditions of the pasture, where there is less moisture in the soil and less protection from the sun.

The researchers intend to continue the project and try to identify those species that have the best chance of surviving the pastures. Different tree seedlings will be compared for survival, growth and drought tolerance to find those that can be best used in reafforestation projects. It seems that returning the degraded land to rain forest will be a long and hard task.

The apparent luxuriance of the tropics continues to deceive newcomers. Contrary to first impressions, the soils of the Amazon Basin have, in general, rather poor agricultural potential. Most of the nutrients are tied up in the vegetation, and they are destroyed when the trees are cut and burned. Rapidly-growing weeds soon invade the cleared land and the great diversity of insects, pests and pathogens makes tropical agriculture a formidable task. After a couple of years, crops can no longer be grown in the cleared areas and the land is usually abandoned or converted to pasture.

The original settlement programme was not successful; it was intended to provide land for impoverished migrants from the northeast of Brazil but, faced with deteriorating crops, soil erosion, disease, and hostility from the local tribes who were

The causes of deforestation are many, though of over-riding importance is the conversion to pasture land for cattle ranching . . .

deprived of their livelihood, the settlers gave up and left the area. They migrated to the Amazon's new cities to try to find work there, returning to the urban slums from which they had originally tried to escape.

Regardless of the early failures, the settlement programmes and development projects still continue, and many are funded by outside agencies. Two of the largest are the Grande Carajás Project, seeking to exploit Brazil's mineral deposits near to the mouth of the Amazon, and the Polonoroeste Project in the state of Rondônia, which is designed to develop the area and relieve population pressure in southern Brazil. In addition, there are plans to build more huge dams which, while generating valuable electricity, will flood large areas of the rain forest and deprive even more Indians of their tribal lands. Often the environmental limitations to the planned projects are known in advance, but are ignored. The projects are designed by governments or businesses as short-term solutions to immediate pressures from their electorates or investors.

Brazil: subsidizing deforestation

The land cleared by the settlers is often bought up by entrepreneurs for cattle ranching. In the past this was encouraged by the Brazilian government through large subsidies and tax incentives. More than US$1 billion has been spent in the last decade to encourage cattle ranching in the Amazon, but it has brought little success and considerable damage to the fragile environment, accounting for perhaps as much as 70 percent of the deforestation in Brazil.

Only 12 trees left – One of the 10 most endangered plants in the world, the Rio Palenque mahogany (*Persea theobromifolia*), is found in a small patch of undisturbed forest in western Ecuador. Only one dozen mature trees exist, but fortunately there are numerous young trees sprouting in the area, and attempts at raising seeds have been successful. The Rio Palenque Science Centre, situated in the area where the mahogany is found, protects a fragment of Ecuador's coastal forest along with its many endemic species.

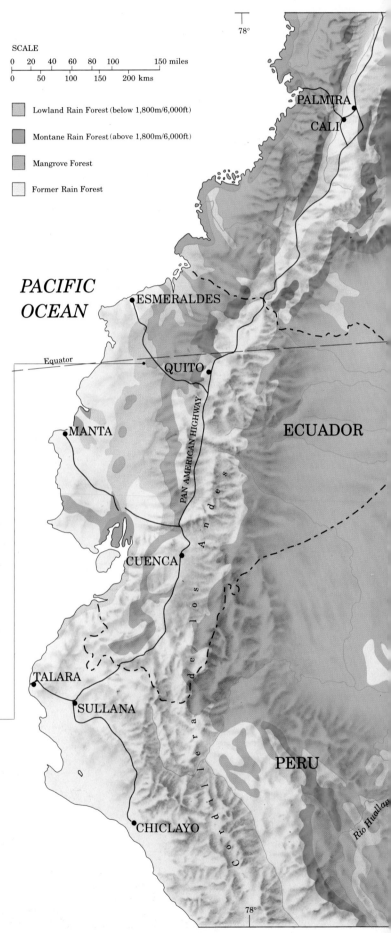

SCALE

| 0 | 20 | 40 | 60 | 80 | 100 | | 150 miles |

| 0 | 50 | 100 | 150 | 200 kms |

Lowland Rain Forest (below 1,800m/6,000ft)

Montane Rain Forest (above 1,800m/6,000ft)

Mangrove Forest

Former Rain Forest

PALMIRA

CALI

PACIFIC OCEAN

ESMERALDES

Equator

QUITO

MANTA

ECUADOR

PAN AMERICAN HIGHWAY

CUENCA

Andes

TALARA

SULLANA

Cordillera de los

PERU

CHICLAYO

Rio Huallaga

78°

78°

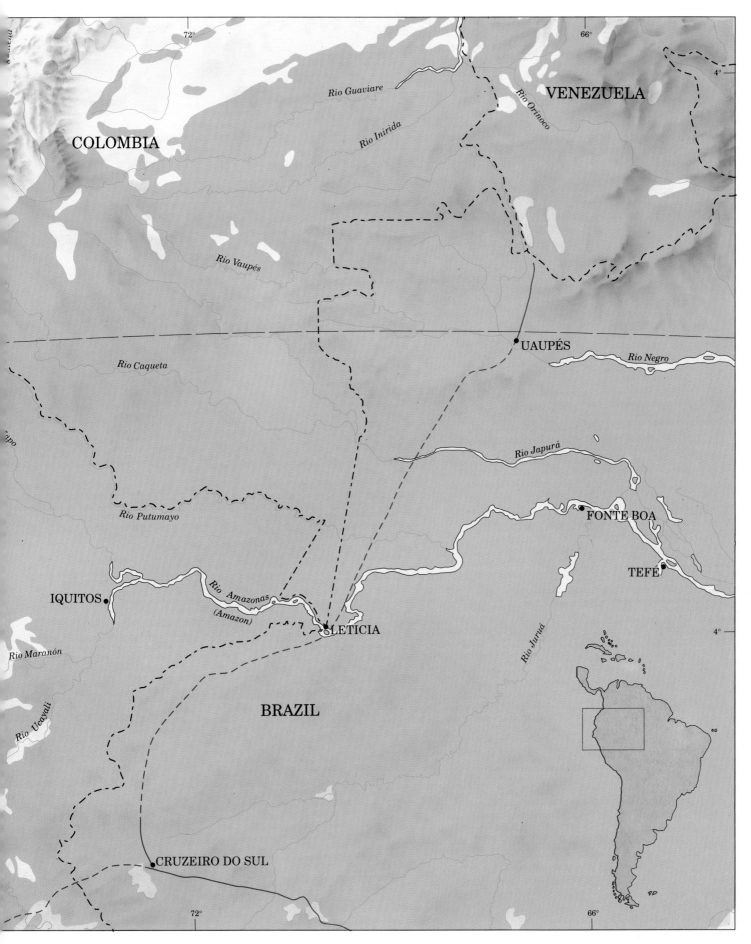

COLOMBIA

Rio Guaviare

Rio Inirida

VENEZUELA

Rio Orinoco

4°

Rio Vaupés

UAUPÉS

Rio Negro

Rio Caqueta

Rio Japurá

FONTE BOA

Rio Putumayo

TEFÉ

IQUITOS

Rio Amazonas
(Amazon)

LETICIA

Rio Juruá

Rio Marañón

4°

Rio Ucayali

BRAZIL

CRUZEIRO DO SUL

72°

66°

The pastures, which initially support one head of cattle per hectare (2.5 acres), rapidly lose their fertility so that within six years more than three hectares (7.5 acres) are needed to support a single animal. Annual meat production on these ranches is as low as 50 kilogrammes per hectare (278 pounds per acre), whereas north European farms can produce 600 kilogrammes per hectare (1.5 tons per acre) within a year, even without the addition of artificial fertilizers.

Both the settlers and cattle ranchers make extensive use of fire either to burn the original forest or to encourage new grass growth on the pastures. Frequently out of control, these fires contribute to the destruction of thousands more square kilometres of rain forest.

Brazil: logging in the future

The Amazon forests seldom contain a high density of a single tree species or species group that could be used to supply large quantities of a single type of wood to either the domestic or international timber markets. As a result, they were long

> ... the Amerindians have a vested interest in conservation as the future of their society depends on [it]. ... There is an Amerindian proverb which illustrates this: "The gods are mighty, but mightier still is the jungle".

considered of little economic value. However, as the supply of timber from other rain forests in Southeast Asia and West Africa diminished, Amazonian species have become more in demand. Brazil's Amazonian log production increased from 4.5 million cubic metres (160 million cubic feet) a year to 19.8 million cubic metres (700 cubic feet) a year between 1975 and 1985.

Since the late 1970s only five species of tree, out of an estimated 1,500 species, have accounted for 90 percent of Brazil's Amazonian timber exports. At present, the selective logging used to extract these species is very wasteful: one tree per hectare (2.5 acres) is extracted, but as loggers move in with roads and skidders they damage or kill more than half of the trees in the area. Improvements in logging techniques will have to be developed to ensure that the destruction caused in Southeast Asia and West Africa is not repeated here.

The northern edge: Venezuela to French Guiana

There is comparatively little deforestation occurring in the countries north of Brazil. Surinam is one of the least densely populated tropical countries in the world, most of its people living along the coast. Only about five percent of the 400,000 inhabitants live in the interior, and they are mainly in villages scattered along three major rivers. This leaves most of the forest uninhabited and undisturbed. The country also has an excellent network of protected areas, including examples of all the major ecosystems, and it plans to expand this still further.

Similarly, in Guyana and French Guiana most of the populace is concentrated in the coastal belt, and there is little exploitation of the forests. Some selective logging is carried out in Guyana, but its impact on the forest is minimal, as is harvesting for charcoal and fuelwood. In Venezuela, logging has been occurring for decades north of the Orinoco River and there is a significant sawmill industry there. The northern forests have also been replaced by crops and livestock. However, the great bulk of the forest lies to the south of the river and here human impact is still

Surinam: setting parrot quotas

Twenty one of Surinam's 30 indigenous parrot species are included in a project set up to control the export of some of the country's wildlife species. The parrots chosen are still abundant in the forests and considered to be pests because of the damage they do to crops.

Having reviewed data on the distribution and status of the parrots (including the three species pictured below) and consulted with the international scientific community, the Nature Protection Division of the Surinam Forest Service (LBB) and the Foundation for Nature Conservation in Surinam (STINASU) have set export quotas for each species. All exporters must belong to the Association of Animal Exporters, and they are required to report on their trapping activities. STINASU biologists monitor the trapping methods used and check that the captured birds are looked after correctly. No bird can leave the country without an export permit and a veterinary certificate. A fee is charged for each shipment and this money covers some of the costs of managing the quota system.

LBB has established a minimum value in US dollars for each species exported. The exporter must receive at least this sum and in dollars. The foreign currency is paid to Surinam's Central Bank, which then pays the exporter in local currency. The quotas can be adjusted if new information suggests this is necessary but, so far, data indicates that exports could be increased with no detrimental effect on the parrot populations. In this way the country is obtaining valuable dollars as well as monitoring the parrot trade and reducing illegal trapping.

Brown-throated parakeet

Blue-yellow macaw

Orange-winged parrot

Sakuddei shaman (*right*) with his brothers

The death of a medicine man, or shaman, has been likened to the burning of a unique library: the knowledge held in his head is not obtainable anywhere else.

Learning from the shamans

Among its many treasures, the Amazon rain forest harbours a plethora of plants that could make a vital contribution to modern agriculture, industry and medicine. Although it is possible for chemical companies to analyze the plants and work out what they might be useful for, the quickest way to get to know the different species and their uses is to ask the local tribespeople. Sadly, though, because more than 90 Amazonian tribes have died out this century, much of the knowledge has already been lost, and the rest is dying along with the Amerindians. The death of a medicine man, or shaman, has been likened to the burning of a unique library: the knowledge held in his head is not obtainable anywhere else.

Tropical plants have already provided many of the world's crops, and the rain forests are certainly a potential source of many more. A possible new, cultivatable species is the uvilla (*Pourouma cecropiaefolia*) from the western Amazon. The fruit from this medium-sized tree is eaten raw by the local Indians or can be made into a wine. The tree is both harvested in the wild and cultivated as a "doorstep" crop. Another potentially valuable species is the pataua palm (*Jessenia bataua*). It produces an oil almost identical to olive oil in its chemical and physical properties, while being 40 percent richer in protein than soyabean oil. Forest plants can also be used to introduce new genetic material into species that are already being used as crops.

Natural pesticides can be extracted from tropical plants that have evolved chemical defences to deter predation by herbivores. An example of this is *Lonchocarpus*, the sap of which is commonly used as a fish poison in South America, and is the source of much of the world's rotenone, an important biodegradable pesticide. Another species that might prove useful as a pesticide is a woody vine, guarana (*Paullinia cupana*), found in central Brazil. This contains three times as much caffeine as coffee and it appears that caffeine can kill or inhibit the growth of many insects.

In the developed countries, annual sales of drugs containing natural plant material is a multi-billion dollar business. Many of these plants were "discovered" through noting for what purpose the local Indians used them. For example, curare, the black resin from a South American tree, is used as an arrow poison by the Amerindians, and now an alkaloid extracted from it is used as a muscle relaxant in surgery. There is undoubtedly a vast storehouse of potentially useful medicinal plants within the forest, and it is the shaman's knowledge that will help the developed world to find them.

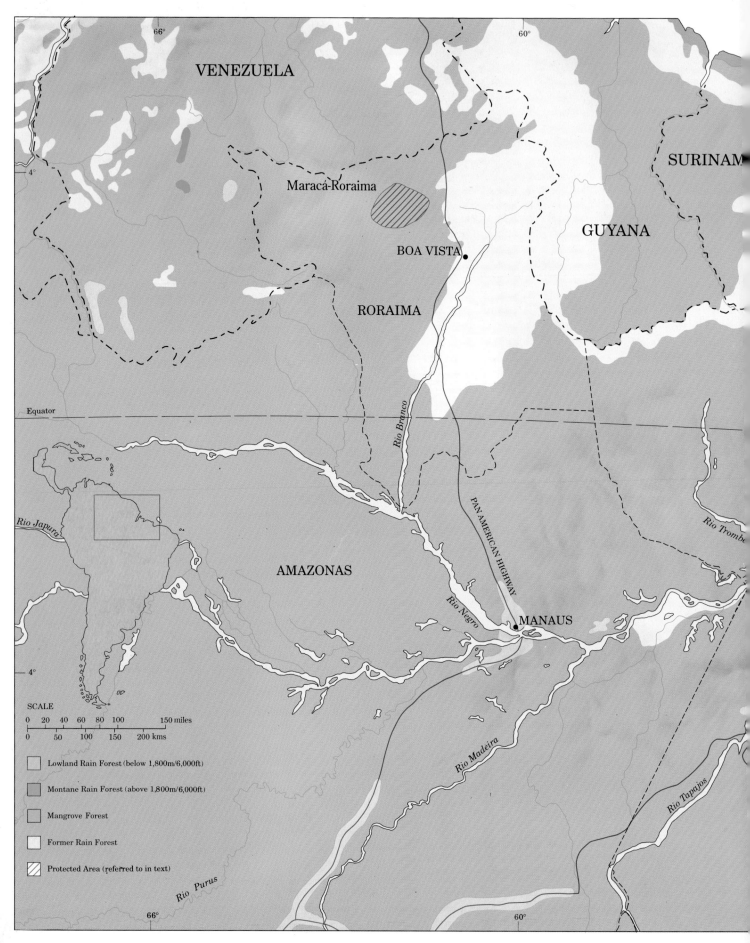

VENEZUELA

Maracá-Roraima

BOA VISTA

RORAIMA

SURINAM

GUYANA

Rio Japura

Rio Branco

Equator

AMAZONAS

PAN AMERICAN HIGHWAY

Rio Negro

Rio Tromb

MANAUS

Rio Madeira

SCALE

0	20	40	60	80	100		150 miles
0		50	100	150		200 kms	

Rio Tapajos

Lowland Rain Forest (below 1,800m/6,000ft)

Montane Rain Forest (above 1,800m/6,000ft)

Mangrove Forest

Former Rain Forest

Protected Area (referred to in text)

Rio Purus

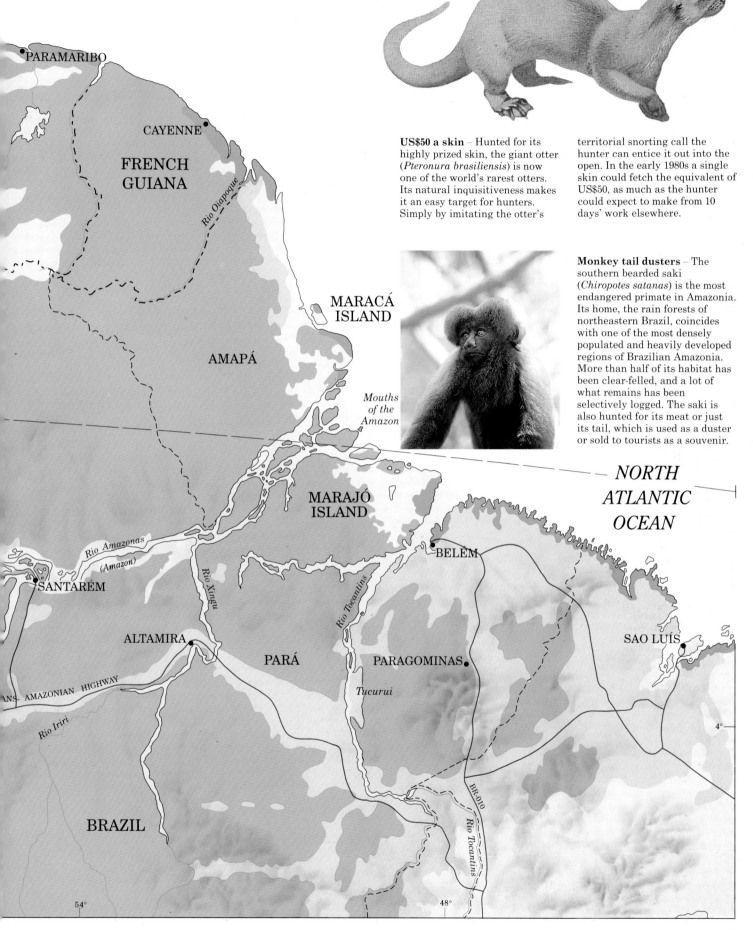

PARAMARIBO

CAYENNE

FRENCH
GUIANA

Rio Oiapoque

MARACÁ
ISLAND

AMAPÁ

Mouths
of the
Amazon

MARAJÓ
ISLAND

NORTH
ATLANTIC
OCEAN

Rio Amazonas
(Amazon)

BELÉM

SANTARÉM

Rio Xingu

Rio Tocantins

ALTAMIRA

PARÁ

PARAGOMINAS

SAO LUÍS

Tucurui

TRANS-AMAZONIAN HIGHWAY

Rio Iriri

4°

BRAZIL

BR-010

Rio Tocantins

54°

48°

US\$50 a skin – Hunted for its highly prized skin, the giant otter (*Pteronura brasiliensis*) is now one of the world's rarest otters. Its natural inquisitiveness makes it an easy target for hunters. Simply by imitating the otter's territorial snorting call the hunter can entice it out into the open. In the early 1980s a single skin could fetch the equivalent of US\$50, as much as the hunter could expect to make from 10 days' work elsewhere.

Monkey tail dusters – The southern bearded saki (*Chiropotes satanas*) is the most endangered primate in Amazonia. Its home, the rain forests of northeastern Brazil, coincides with one of the most densely populated and heavily developed regions of Brazilian Amazonia. More than half of its habitat has been clear-felled, and a lot of what remains has been selectively logged. The saki is also hunted for its meat or just its tail, which is used as a duster or sold to tourists as a souvenir.

slight. There is a total ban on logging in the state of Bolivar, and the Venezuelan Government has decreed that 60 percent of the country's forests are to be left undisturbed.

Colombia

As in Brazil, cattle ranching, government-sponsored settlement projects and shifting cultivation are the main causes of deforestation. In 1988, it was estimated that as much as 6,000 square kilometres (2,300 square miles) of forest was being cut each year, mainly on the Andean slopes. To the east, Colombia's Amazonian forests are still virtually untouched. Colombia has recently developed a Forestry Action Plan, the objectives of which are to preserve its Andean forests while raising the living standards of the poorest groups in the country, and increasing the contribution made by the forests to the country's economy. In addition, President Virgilio Barco has been active in ensuring that Colombia's indigenous Amazonian peoples have control of their land. Nearly 200,000 square kilometres (77,200 square miles) of the rain forest, about half that in the country, are now under the control of the Indians. It is theirs in perpetuity and cannot under present law be sold or exchanged.

Ecuador

Population growth is higher in Ecuador than in any other South American country, and this is one of the main causes of deforestation there. Most of Ecuador's lowland forests to the west of the Andes have been removed in the last 30 years. The first road penetrated that area in 1960 and within 12 years the original forest was almost completely converted to plantations of cash crops, such as oil palm and banana. These lowland forests were some of the richest, and are now some of the most threatened, in the world. The FAO estimates that 3,400 square kilometres (1,300 square miles) is lost each year.

To the east of the Andes, in Ecuador's Amazonian forests, oil exploration is, both directly and indirectly, the principal cause of destruction. Exploration has been allowed in some reserves but, more importantly, extensive road building by the oil companies has enabled settlers to move in and clear the trees regardless of whether they were in protected areas or tribal lands. Pollution associated with the oil exploration is also causing problems in some of the rivers and streams. The country is very aware of the need to conserve its forests; it is one of the four Amazonian countries to belong to the International Tropical Timber Organization (ITTO), and IUCN has been invited to set up its first South American office there. Ecuador is also participating in a project to raise money for conservation by means of an international debt swap, as are Bolivia and other countries outside South America.

Peru

The principal threat to Peru's rain forests is from agricultural settlement. Deforestation followed by overgrazing has left many of the Andean slopes bare of vegetation and highly susceptible to wind and water erosion. As a result, the country's reservoirs and irrigation systems are becoming increasingly silted up. In 1981, Peru was estimated to have 700,000 square kilometres (265,000 square miles) of tropical forest left, more than any other South American country apart from Brazil. Today, deforestation is estimated to average around 2,600 square kilometres (1,000 square miles) a year, most as a result of the growth in population and the consequent need for land, but large areas are also cleared to grow cash crops.

Coca, used in the manufacture of cocaine, is an increasingly important crop in Peru; no other crop apart from coffee is planted so extensively. For instance, in the Huallaga Valley in the eastern Andes between 1,600 and 3,800 square kilometres (620 and 1,470 square miles) are planted with coca. Its cultivation on steep hillsides has led to massive erosion, and the herbicides, including Agent Orange, used to control the weeds on the plantations may well cause pollution problems in the area. The camps, processing

Once isolated, small forest fragments soon degrade.

Islands in the jungle

In 1979, a unique long-term experiment was set up to study the effects of dividing the forest up into small fragments, or islands. Such islands are typically formed when cattle ranchers clear an area for pasture. Situated in the rain forest 105 kilometres (65 miles) north of Manaus in Brazil, the project is being run by the World Wide Fund for Nature and Brazil's National Institute for Amazon Research.

First 23 "reserves", varying in size from one hectare to 100 square kilometres (2.5 acres to 40 square miles), were marked out and their flora and fauna studied to obtain a "before" picture of the habitat. The forest islands were then formed and any changes noted.

The greatest changes have been recorded in the smaller fragments. In one of the one-hectare islands, only 18 of the original 39 bird species remained after a year of isolation. The red-handed tamarin (*Saguinas midas*), which was abundant in the undisturbed forest, was absent from both one- and ten-hectare islands. In addition, some curious interrelationships between species have been discovered. White-lipped peccaries (*Tayassu pecari*), a type of pig, rapidly disappeared from the smaller reserves. Soon after, three frog species could not be found. It turned out that they were dependent on the pig's wallows for breeding grounds. No peccaries no wallows, and therefore no frogs.

Effects on the vegetation itself have been quite dramatic. Adult trees die faster, weeds invade the edges of the forest islands and hot, dry winds reach into the interior, blowing over trees and changing the temperature and humidity of the forest floor, upsetting the ecosystem. All of the vegetation in the one- and ten-hectare plots was degraded in this way, while only the core of the 100-hectare island remained unaffected.

A narrow forest "bridge", only 100 metres (110 yards) wide, between an island and the undisturbed forest considerably increased the number of species that remained in the island. In the opposite way, a narrow road cut through the undisturbed forest acts as an insurmountable barrier to many species. Much remains to be learned, but it is hoped that the project will help in the design and management of rain forest reserves.

So long as only small areas are mined at any one time, reafforestation does not take too long.

The greening of a bauxite mine

Just north of the Amazon, at Mineraçao Rio Norte on the Trombetas River, there is a large opencast bauxite (aluminium ore) mine, situated in the middle of what used to be completely undisturbed rain forest. To extract the ore, 70 hectares (173 acres) of forest are cleared each year. But after the bauxite has been removed, the mining company is helping the forest to regenerate.

To clear the land for mining all the trees in the area are cut and burned, and 15 centimetres (six inches) of topsoil is removed and stockpiled for later use. The bauxite is then mined, the earth is replaced, levelled, covered with the carefully-saved topsoil and reafforestation begins. About 90 native species and 12 exotic species are used for replanting and many others seed themselves naturally.

Within 18 months there is a lush growth of vegetation over the mined area. Of course it takes several decades before the area begins to resemble its surroundings.

If, as the company is requesting, the Brazilian Environmental Institute declares the mine and its surrounds a conservation area, the company will then take official responsibility for protecting the forest. Only ten percent of the area will be mined and reafforested, the rest will remain untouched to provide a seed bank for the regenerating areas. Hunting is already forbidden in the region and this is enforced to the extent of sacking staff found killing any animals. The company is also supporting an environmental education programme that is intended to increase local knowledge and concern about the forest.

Exploring for oil

In the mid-1960s oil was discovered in the east of Ecuador. Already 6,300 square kilometres (2,400 square miles) of forest have been exploited, and a further 3,000 square kilometres (1,200 square miles) are now being explored for more oil. Although, with careful management, funds obtained from oil resources can be successfully channelled into forest conservation (as has happened in Brunei in Southeast Asia), Ecuador has not fared so well.

Typically, during the exploration stage, a grid of trails one kilometre (0.6 mile) apart is constructed, and helicopter landing sites are cleared along these, also at one-kilometre intervals. Explosives are then detonated at 100-metre (110-yard) intervals to generate sound waves for seismic analysis. The exploration itself does very little damage, and the affected areas are quickly recolonized by the forest on abandonment. Exploratory wells are then dug. Two to five hectares (five to twelve acres) of forest are completely cleared to make way for each well, and a further 10 to 15 hectares (25 to 37 acres) are disturbed to make boards for the drilling platforms. Once oil has been discovered, unless due caution is exercised by the oil company, there is a danger of pollution from waste products such as sulphates, cyanides, mercury and contaminated water.

In Ecuador, the Cuyabeno Wildlife Reserve had a pipeline built through it which led to an invasion of settlers, and the Yasuni Wildlife Park is now threatened. The designation of the Yasuni Park as a biosphere reserve and a world centre for plant diversity appears not to have been enough to save it. Oil wells have already been drilled within the park and there are plans to build a road through it to enable the oil company to lay a pipeline. The resulting invasion of settlers will probably spell the end of the nomadic Waorani Indians, a tribe that until now has avoided all contact with outsiders.

Oil revenues can be used to protect the rain forest.

SCALE

0 20 40 60 80 100   150 miles

0   50   100   150   200 kms

☐ Lowland Rain Forest (below 1,800m/6,000ft)

■ Montane Rain Forest (above 1,800m/6,000ft)

☐ Former Rain Forest

CHICLAYO

TRUJILLO

PACIFIC OCEAN

LIMA

Harvesting the forest, strip by strip

Approximately 6,000 people live in the Palcazu Valley in Peru, of which around half are Amuesha Indians. Although the valley retains about three-quarters of its original forest, there has been considerable clearance along the rivers. The Palcazu Project has been set up to manage the forests to provide the local people with economic independence, without destroying their land.

The management plan divides the lower valley into three categories: land that can support some agriculture and grazing (35 percent); land that should remain as protected forest (18 percent); and forested land that should be harvested (47 percent). The Palcazu system involves harvesting small strips of forest, 20–30 metres (22–33 yards) wide and of variable length, in 30- to 40-year rotations. Each harvested strip must be at least 200 metres (220 yards) from those cleared in previous years and for each block of five strips, a section of forest is left untouched. In effect these strips mimic the small natural clearings that are created when a large tree falls.

During the harvest, major branches and trunks are removed, and small branches and foliage are left behind to provide nutrients for the soil. Oxen are used to take the wood out of the forest, a method both cheaper and much less damaging to the soil than using machinery. The harvested wood is sawn into posts and logs or converted to charcoal at a small, cooperative-owned sawmill. Potential net profits may be as much as US$3,500 per hectare (2.5 acres). In addition, orchids and other ornamental plants can be collected and sold before the forest strip is cleared, as can medicinal and edible plants, and fibres for making baskets, nets and brushes.

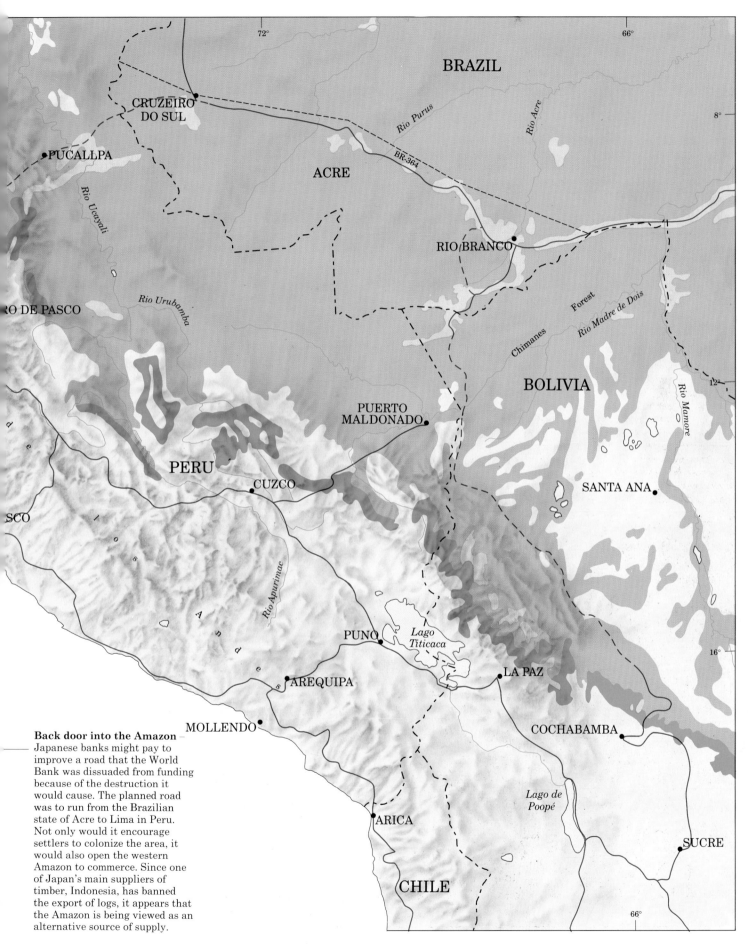

BRAZIL

CRUZEIRO
DO SUL

PUCALLPA

Rio Purus

Rio Acre

ACRE

BR-364

Rio Ucayali

RIO BRANCO

Rio Urubamba

O DE PASCO

Forest

Rio Madre de Dois

Chimanes

Rio Mamore

BOLIVIA

PUERTO
MALDONADO

PERU

SANTA ANA

CUZCO

Rio Apurimac

A
n
d
e
s

SCO

PUNO

Lago
Titicaca

LA PAZ

AREQUIPA

MOLLENDO

COCHABAMBA

Lago de
Poopé

Back door into the Amazon —
Japanese banks might pay to
improve a road that the World
Bank was dissuaded from funding
because of the destruction it
would cause. The planned road
was to run from the Brazilian
state of Acre to Lima in Peru.
Not only would it encourage
settlers to colonize the area, it
would also open the western
Amazon to commerce. Since one
of Japan's main suppliers of
timber, Indonesia, has banned
the export of logs, it appears that
the Amazon is being viewed as an
alternative source of supply.

ARICA

SUCRE

CHILE

laboratories and landing-strips associated with this industry account for further deforestation and pollution.

The refinement of the coca leaves, first into coca base then into coca paste, occurs in Peru. The final processing into white cocaine powder is usually carried out in Colombia. This refining process involves the use of many different chemicals, such as sulphuric acid, acetone, toluene and lime, and vast amounts of each of these end up dumped in the watershed of the Huallaga Valley. Because it is not possible to keep law and order in the areas that are under the control of the drug traffickers, logging, hunting and fishing goes on unchecked.

Perhaps fortunately, there are still few roads to the east of the Peruvian Andes. The major means of transport is the extensive river system. As a result, logging is not a great threat in Peru and it is still difficult for settlers to move in and deforest large areas of the country. This might change if the Japanese government funds a road from Lima, over the Andes, into Brazil.

Bolivia

Deforestation in Bolivia is relatively minor, estimated by the FAO to be proceeding at a rate of 870 square kilometres (335 square miles) a year. Shifting cultivation is not a major threat at present because the human population of 18.3 million is comparatively small for a country the size of Bolivia. But the population is growing at a rate of 2.76 percent a year, and Bolivia is one of the few South American countries where the growth rate from 1985–1990 has remained as high as that for 1965–1970, so it seems likely that increasing pressure will be put on Bolivia's forests.

The lowland moist forests have been subject to little exploitation as yet. Logging is very restricted, being mostly confined to two types of tree, *Virola* sp. and *Cedrela* sp., in the seasonally

---

... the death of Chico Mendes has led, finally, to international acknowledgment of the harm being done by the cattle ranchers and development projects.

---

flooded forests. However, it has led to the near commercial extinction of mahogany (*Swietenia macrophylla*) in the Santa Cruz area. Following the opening of new roads, the department of Beni in the north of the country is now being developed for industrial timber production, the extraction of mahogany increasing sixfold in the ten years between 1977 and 1987.

Bolivia, along with Brazil, Ecuador and Peru, belongs to the ITTO. One of the major aims of this organization is to improve forest management for sustainable timber production. A project supported by the ITTO has recently been set up in the Chimanes forests in Beni, which aims both to conserve the area and use its forests on a sustainable basis.

Future conservation

The peoples in many of the Amazonian countries are becoming increasingly aware of the need to protect their environment. In Brazil, large numbers of pressure groups have been formed to convince the government that something must be done. The Brazilian rubber tappers are determined that they should be allowed to continue to use the forest, and the death of Chico Mendes has led, finally, to international acknowledgment of the harm being done by the cattle ranchers and development projects. The tax incentives for cattle ranchers have been stopped, and there is a ban on the export of unworked timber. International aid agencies and bodies such as the World Bank are now more careful that the money they lend to South American countries is not used in a way that is detrimental to the rain forest environment and its peoples.

Rubber tapping (*above, below right*) – The forests in the Brazilian state of Acre are still virtually intact, only 4% having been lost so far. But the rate of deforestation is increasing as cattle ranchers move in and clear the trees to make way for pastures. In conflict with these ranchers are the rubber tappers who have been in the area since the mid-1800s. Some 500,000 people earn their living from collecting the latex from rubber trees that grow wild in the forest (*below right*). The rubber tappers do not have title to the forests they harvest, so they have organized a series of cooperatives aimed at gaining legal guarantees for maintaining their right to use the forest. Francisco "Chico" Mendes Filho (*above*), a tapper himself, led meetings of the National Council of Rubber Tappers in its attempts to protect the interests of its members. Obviously, cattle ranching is not compatible with the "extractive reserves" that the rubber tappers are demanding, and it is this state of affairs that led to the murder of Chico Mendes. Mendes did not set out to save the Amazon as such, but to improve the lot of the rubber tappers, and this meant preventing deforestation. He organized non-violent protests, forming the rubber tappers and their families into a human barrier in the way of advancing chainsaws, and he then attempted to negotiate with the loggers to prevent the destruction of yet more land. Mendes survived 6 attempts on his life but in December 1988 he was shot and killed as he left his house. He has become a hero to environmentalists because he died trying to show that it is possible to earn a living from the forest without destroying it. Indeed, since his death, a scientific study has shown how

right he was. The income from fruits, latex and other forest products harvested on a sustainable basis can exceed that obtained by logging or ranching where conditions, including accessible markets and local demand, are favourable. Even more important, just as Mendes wanted, it is the local people who benefit from the forest used in this way, not foreign logging companies or rich cattle-ranching entrepreneurs.

Rural slums (*above*) – Brazil's Northwest Regional Development Project, or Polonoroeste, was designed, with the assistance of the World Bank, to accelerate the economic development of the state of Rondônia. The centrepiece of the programme, the construction of a road (BR364) from Cuiabá to Pôrto Velho, was completed in September 1984. In the first 3 months of 1985, 15,000 families entered Rondônia, and by July of that year 50,000 families were awaiting settlement, and many more have followed since. On arrival, the settlers often build temporary shelters beside the road. They then cut down the forest and grow annual crops on the nutrient-poor soil for 2 or 3 years, before the land is bought up by entrepreneurs for cattle ranching. It is then converted to pasture and within a decade the already degraded soil is useless. For the settlers, the effect is of having moved from urban slums to rural slums, but they are still better off than many of the local Indians: removed from their native homes, clashing violently with the newcomers and dying from introduced diseases, the Indians are unable to survive the changes. In 1980, 97% of Rondônia was forested, now that figure is down to 80% and more is being cleared every day.

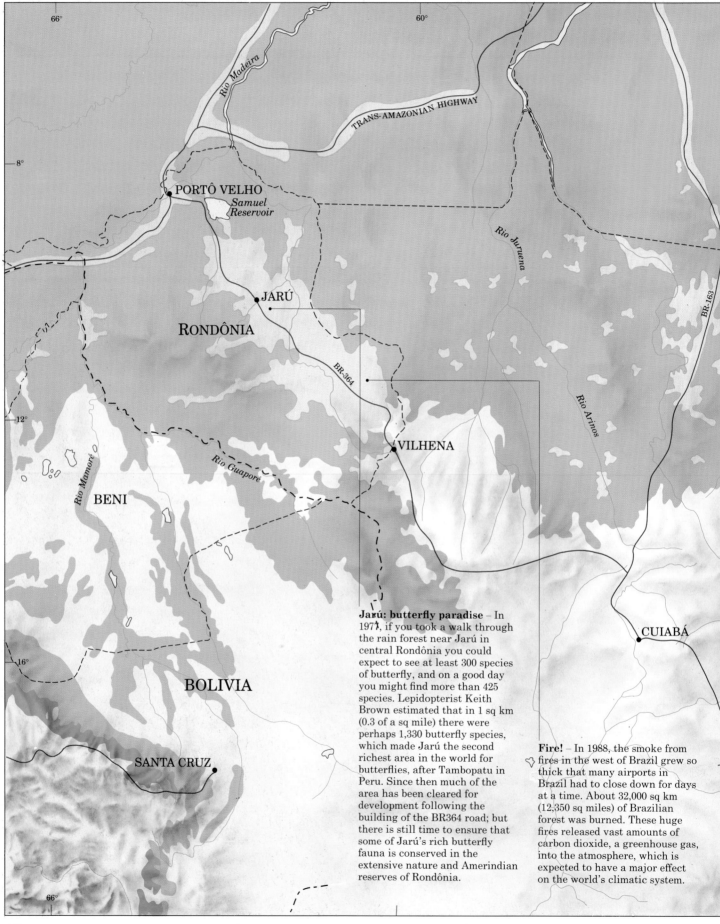

66°   60°

Rio Madeira

TRANS-AMAZONIAN HIGHWAY

8°

● **PORTÔ VELHO**
*Samuel
Reservoir*

Rio Juruena

● **JARÚ**

RONDÔNIA

BR-364

BR-163

Rio Arinos

12°

Rio Mamoré

Rio Guaporé

● **VILHENA**

BENI

16°

BOLIVIA

● **CUIABÁ**

● **SANTA CRUZ**

Jarú: butterfly paradise – In 1971, if you took a walk through the rain forest near Jarú in central Rondônia you could expect to see at least 300 species of butterfly, and on a good day you might find more than 425 species. Lepidopterist Keith Brown estimated that in 1 sq km (0.3 of a sq mile) there were perhaps 1,330 butterfly species, which made Jarú the second richest area in the world for butterflies, after Tambopatu in Peru. Since then much of the area has been cleared for development following the building of the BR364 road; but there is still time to ensure that some of Jarú's rich butterfly fauna is conserved in the extensive nature and Amerindian reserves of Rondônia.

Fire! – In 1988, the smoke from fires in the west of Brazil grew so thick that many airports in Brazil had to close down for days at a time. About 32,000 sq km (12,350 sq miles) of Brazilian forest was burned. These huge fires released vast amounts of carbon dioxide, a greenhouse gas, into the atmosphere, which is expected to have a major effect on the world's climatic system.

66°

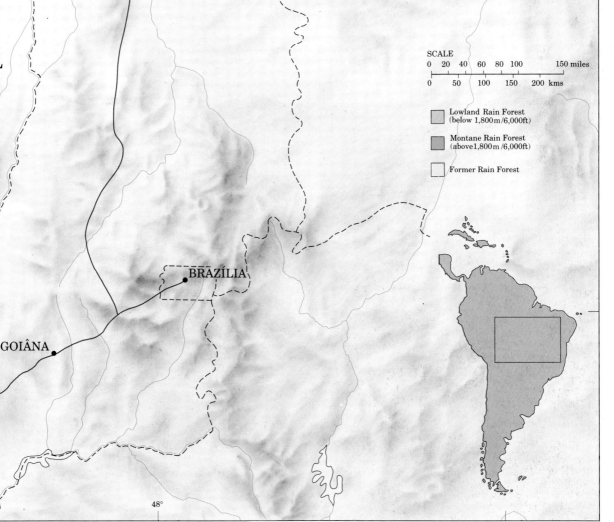

Most of the Amazonian countries have set up more protected areas, on paper at least. Instead of providing money to exploit and extract the wealth of the forest, as has happened previously, the developed countries need to ensure that their money is used to investigate the biological and physical basis of the forests' wealth, including research into the possible uses of forest plants in industry, agriculture and medicine. Some areas must be left undisturbed and this will mean more money is needed to ensure that they are adequately protected.

Another possibility is the development of the land outside the rain forest. In Brazil, there are fewer environmental risks and many major economic advantages in promoting development in the large areas of savanna and scrub (cerrado) in the south of the country. Here resources can be used to improve agricultural land that is already in production.

Undoubtedly most knowledgable and skilled in the subsistence use of the Amazonian forest are the local inhabitants, the Amerindians, and it is they who have often suffered from the development of the Amazon. Unlike the settlers, the Amerindians are not exploiting the natural resources on a short-term, high-profit basis. They have a vested interest in conservation because the future of their society depends on the survival of the forests. There is an Amerindian proverb that illustrates this: "The gods are mighty, but mightier still is the jungle". The people's interests must be considered to have high priority in the Amazon Basin, and working with them, rather than against them, will surely help outsiders, whether scientists or settlers, in understanding some of the complexities of the forest.

SCALE

| 0 | 20 | 40 | 60 | 80 | 100 | | 150 miles |

| 0 | 50 | 100 | 150 | 200 kms |

Lowland Rain Forest (below 1,800m /6,000ft)

Montane Rain Forest (above 1,800m /6,000ft)

Former Rain Forest

The native people of the Amazon Basin

A Portuguese expedition to the Amazon Basin in 1637 reported the Indians to be "so numberless that if a dart were to fall from the air, it would strike the head of an Indian and not fall on the ground." The report referred to the Amazonian floodplain, in the regions known as *várzeas* (see pages 20–21), where the fertile soil then supported a thriving population of agriculturalists. Permanent settlements lined the riverbanks, and these were divided into provinces, ruled by all-powerful chieftains. Following the development of a variety of manioc (cassava) that matured in six months rather than the usual twelve, the *várzea* inhabitants grew all the food they needed during the nine months of low water. Food was stored for the three months of the year when the *várzea* was inundated.

A century after the expedition of 1637, the tribes of the *várzeas* had gone. Many had been taken as slaves by the European invaders. Many more died during epidemics of influenza and other new diseases, to which the Amerindians had little resistance. A thriving, well organized civilization had been wiped out in just a hundred years.

Much the same sequence of events has been acted out over the whole of Amazonia, albeit more slowly, and the final stages of this tragic annihilation seem now to be taking place. The inhabitants of the *terra firme* forests, more dispersed and less accessible than those of the floodplain, have survived in significant numbers, although their persecution has been relentless, and many tribes have disappeared altogether. Those still living in the forest are generally in remote locations. The survivors also tend to be the more ferocious tribes, whose intrinsic suspicion of outsiders, and aggressive behaviour towards anyone entering their territory, has protected them. Considering the way in which the outside world has treated the indigenous people of Amazonia – many of whom treated the newcomers with generous hospitality – such an attitude is entirely justified.

The catalogue of persecution and injustice that has afflicted these people is almost endless. At one time, diseases such as smallpox were deliberately introduced, impregnated into blankets. During the rubber boom of 1840–1912, whole tribes were enslaved by rubber producers in an iniquitous system of debt bondage. People of European descent have long regarded the Indians as "animals" and have shot them for no reason – something that still occurs today. Their lands have been seized and large populations massacred or shifted to other, less suitable areas of the forest.

In the past 20 years, the pressure on the native people of Amazonia has increased enormously, as modern technology, compelled by new political drives for expansion, opens up the remaining areas of rain forest to development. As in the past, previously isolated populations succumb to epidemics of European diseases. Conflicts with gold miners and cattle ranchers have resulted in many more deaths, and the poisoning of streams with mercury (used to separate out the gold) has caused illness among groups such as the Kayapó. The building of roads and dams still displaces large populations, and the relentless burning and bulldozing of the forest destroys the very basis of many of the Amazonian tribes' existence.

Missionaries, principally Roman Catholic, have been in Amazonia for centuries, and in the past have had both good and bad effects, with the benefits probably outweighing the drawbacks. Today, a new breed of fundamentalist zealots, with an uncompromising approach to conversion, and an apparent lack of concern for the earthly fate of the Amerindians, is operating in Amazonia to the physical and cultural detriment of a great number of ethnic communities.

Against this background, there are some glimmers of hope. The native people of Amazonia have found a new self-confidence and pride in their Amerindian identity. They have established bonds of solidarity and cooperation between different villages and tribes, where previously there was mutual hostility. Some are learning Portuguese and international awareness of the problems they face has given them greater strength and increased their political power.

Whether this new awareness can lead to the survival of the Amerindians, and that of their forests, remains to be seen. The forces ranged against them are many, and time is running out.

Body decoration (*left, above*) – The designs used by the members of this tribe of Amerindians living near the Xingú River in the Amazon Basin follow an unbroken cultural tradition of centuries.

A Yanomami headman (*right*) lives communally with other members of his tribe in a large palm-thatched hut or *shabon*.

The Yanomami

More than 20,000 Yanomami (or Yanomamo) still live in the highland rain forests around the border between Brazil and Venezuela. They are the largest Amerindian group in Amazonia that is still following a traditional lifestyle, although this has been modified by contacts with outside. Like most Amazonian tribes, the Yanomami combine shifting cultivation (see page 93) with hunting and the gathering of forest foods. They are unusual in using the plantain (a starchy type of banana) as their staple crop – it supplies about 70 percent of their food. There are many Yanomami dialects, but they preserve a traditional form of language for ceremonial use, and this can be understood by all Yanomami. The Yanomami's territory covers about 40,000 square kilometres (15,400 square miles). In the past, warfare between different villages was common especially in the central part of the territory. Here villages were highly fortified, while on the outer edges there was less tension and a village could always decide to move outwards into virgin territory if

threatened by aggressors. Almost all the villages moved regularly to escape enemies or form new alliances.

This state of affairs served to keep the settlements from exhausting the soil in any one area and to prevent excessive population growth. It also bred a fierce isolationism in the Yanomami warriors which served them well in their contacts with Europeans, and they remained relatively untroubled by the outside world until the early 1970s. Government plans to build Highway BR210 across their territory brought road-building gangs and the inevitable devastating epidemics. This was followed in 1985 by an attempt to bring in several thousand miners by air, an effort backed by local politicians but finally thwarted by the federal government. Incursions by gold miners, and new epidemics, continue. United by the common threat to their survival, the Yanomami are now demanding protection for their lands. These demands have so far been ignored by the authorities and the Yanomami's future remains uncertain.

The Atlantic coast of Brazil

SCALE

0	20	40	60	80	100		150 miles
0		50	100		150		200 kms

Lowland Rain Forest (below 910m/3,000ft)

Mangrove Forest

Former Rain Forest

Protected Area (referred to in text)

ATLANTIC OCEAN

SOS Atlantic Forest Foundation

An influential group of Brazilian scientists, businessmen and journalists joined forces at the end of 1986 to form the Fundação SOS Mata Atlantica (SOS Atlantic Forest Foundation). Since then, numerous individuals and organizations have devoted their time and money to trying to save the last vestiges of the Atlantic forests. The nation's media have provided widespread coverage of the movement, including a three-week television campaign in December 1987. A Brazilian advertising agency, DPZ, gave its services free and has developed a logo for the campaign – the green Brazilian flag with a corner torn away – which symbolizes the destruction of the country's green forests. The Foundation has been allowed access to satellite photographs of the Atlantic forest area so that it can work out the rate and pattern of deforestation and the impact of subsequent erosion and flooding. The group has several priorities: to set up and manage specific conservation sites; to develop sustainable ways of using natural resources; to increase environmental awareness throughout the country; and to encourage research into the forests. The appearance of the SOS Atlantic Forest Foundation reflects changing attitudes throughout Brazil. More than 500 new pressure groups have been formed in the last five years.

Mention deforestation in Brazil and most people immediately think of the destruction of the Amazonian rain forests. However, in the east of the country there are other, entirely different forests or, more accurately, a few remnant patches of those forests, which are in much greater danger of disappearing. The Atlantic forests used to cover about one million square kilometres (385,000 square miles), stretching from the state of Rio Grande do Norte at the easternmost tip of South America down as far as Rio Grande do Sul, the southernmost state in Brazil, in a strip ranging from several to 160 kilometres (100 miles) wide.

These forests were some of the finest in the world and even now, although they are reduced to a mere one to five percent of their original extent, they contain an incredible richness and diversity of life. The state of the lowland forests is particularly precarious, whereas the montane forests on the slopes of the Serra do Mar have been protected to a certain extent by the steepness of the terrain. Little clearance has occurred above 1,000 metres (3,300 feet); in the state of Rio de Janeiro it is illegal to clear forests above this height, but it is not always possible to enforce the law.

Early visitors, including the young Charles Darwin, were overwhelmed by the luxuriant high Atlantic forests, festooned in orchids and bromeliads and full of a variety of birds, insects, mammals and other animals. Indeed, the forests are still of special interest because of the large number of unique species that they contain. For instance, there are an incredible 2,124 species of butterfly: two-thirds of all Brazil's butterflies, and one-eighth of the world's butterfly fauna. Of these, 913 are endemic. Similarly, 17 of the 21 primate species in the Atlantic forests are unique to the region, as are more than half of the tree species. As might be expected, many of these are endangered and it is quite possible that some have already become extinct. This is possibly the fate of one of the smallest birds in the region, the kinglet calyptura (*Calyptura cristata*), which has not been seen this century.

The arrival of the Europeans

The area occupied by the Atlantic forests has been inhabited for many thousands of years. The Amerindians, who were hunter-gatherers, arrived around 10,000 years ago, but they probably had little impact on the forest. About 1,500 years ago, they were driven from the forests to the less productive highlands of the interior by the Tupi-Guaraní. These Indians were shifting cultivators, planting manioc and other crops in place of the forest trees. However, they too were not an important agent in the deforestation; it is estimated that each family cleared only one hectare (2.5 acres) of forest a year.

As a result of its coastal location, the Atlantic forest area was the first place in South America to be colonized by Europeans, in 1500. This is when the destruction began. Ports and trading posts were set up all along the coast, and these provided the first bases for exploration inland.

Brazil's trees – When the fruit of a Brazil nut tree (*Bertholletia excelsa*) ripens it splits open to reveal the hard, edible nuts within (*above*). Brazil nut trees are found along the banks of the Amazon and Orinoco Rivers, and their fruits are one of Brazil's most important rain forest products. In contrast, the tree from which Brazil got its name, the American Brazilwood (*Caesalpinia echinata*), was once used in the production of red and orange dyes, which were exported to Europe and North America in large quantities during the 18th century. By the 19th century, supplies had been severely reduced and now the tree is of no commercial importance. The American Brazilwood is even difficult to find in its native habitat.

Only enough for veneer – The tree that provides the most valuable wood in all of Brazil is now on the endangered list. The Brazilian rosewood (*Dalbergia nigra*) has long been in demand for furniture-making, but it has been so intensively logged that supplies can no longer meet the demand, even though the timber is now ordinarily used only as a thin veneer to cover less attractive woods. There are no commercial plantations of this species, mainly because it was thought to be too slow-growing. However, research has begun into the possibility of improving cultivation techniques.

Butterflies everywhere –
With more than 2,000 species of
butterfly, including *Dismorphia
amphione* (*left*), *Callithea
leprieuri* (*top*), and *Callithea
sapphira* (*above*), the Atlantic
coast forests are a veritable
butterfly paradise. But along
with the rain forests, many of
these butterflies are in danger
of extinction.

The Pataxo (*left*) – Monte
Pascoal, a national park in the
state of Bahia, was established in
1961. At that time, both Pataxo
Indians and other Brazilians
were living in the area. The
latter were moved and the
Brazilian agency concerned with
the rights of the Indians
attempted to have the area
declared Pataxo Indian land. But
in response, the agency
responsible for forests and
national parks decreed that the
Indians must also be moved. In
August 1980, as a result of
increasing pressure from the
Pataxo, some 85 sq km (33 sq
miles) of the park were returned
to them. This land, largely
forested at the time it was
handed over to the Indians, is
now completely cleared, and
much of it is abandoned pasture
land. Although before 1961 this
area had been occupied by the
Pataxo and provided them with a
sustainable livelihood for more
than a hundred years, it appears
that it can no longer do so. One
of the main reasons for this is
probably the pressure brought to

bear on them to change the way
in which they harvest the forest.
It has been claimed that local
timber merchants have provided
chainsaws and have encouraged
the Pataxo to cut the timber in
the park. In the meantime, the
conflict between the agencies
continues. The Pataxo are
demanding that the whole Monte
Pascoal National Park be given
to them, and yet that seems to
spell certain disaster for the
remaining 140 sq km (54 sq miles)
of one of Brazil's most visited
national parks.

Red howler monkey

Woolly spider monkey

Dusky titti

Pale-fronted capuchin

Common marmoset

Golden lion tamarin

Atlantic forest primates

The Atlantic forest region is home to six primate genera (*Callithrix*, *Leontopithecus*, *Callicebus*, *Cebus*, *Alouatta* and *Brachyteles*), including as many as 21 species and subspecies of monkey. More than 80 percent of these primates are found only in the Atlantic forests and, consequently, at least 13 are considered to be endangered; another three are vulnerable. Several of the endangered species are on the verge of extinction, only a few hundred members remaining in the wild. Two of these, the golden lion tamarin (*Leontopithecus rosalia*) and the muriqui or woolly spider monkey (*Brachyteles arachnoides*), are members of the two endemic Atlantic forest primate genera, and their situation can be taken as indicative of what is happening in the region as a whole.

The muriqui is the largest primate in South America. There were probably as many as 400,000 of these delightful creatures in the Atlantic forests when Europeans reached Brazil in the sixteenth century. Indeed, they were so abundant that some early expeditions were able to live entirely on their meat. By the early 1970s, there were perhaps 3,000 individuals remaining, but now only 11 populations are thought to survive, totalling a mere 400 animals. Although the muriqui has long been a target for hunters and is still poached for food, sport or to obtain an infant as a pet, the main factor in its decline has been habitat destruction.

As the forest area becomes smaller and more fragmented, the muriqui is usually the first primate to disappear; it is more demanding than the other primates in its choice of habitat; less able to use the remaining forest fragments; and slower to re-establish itself in protected

areas. Fortunately, the precarious state of this unique primate has been recognized in Brazil; it has become the national symbol for conservation. There is now a concerted effort to save it and its forest home from extinction.

The story of the golden lion tamarin and of the other two species in the same genus, the golden-headed lion tamarin (*Leontopithecus chrysomelas*) and the golden-rumped lion tamarin (*L. chrysopygus*), is similar to that of the muriqui. They too are disappearing as the forests are destroyed. The strikingly-coloured golden lion tamarin has always been restricted to the coastal lowlands of the comparatively small state of Rio de Janeiro. Only two wild populations are now thought to exist; one of them in the Poço das Antas Biological Reserve, which was established in 1974 mainly for its protection; and the other in a stretch of forest along the coast from the mouth of the Rio São João. Between these two areas, there are probably as few as 250 individuals left.

The golden lion tamarin has long been kept as a pet, even by eighteenth-century European royalty, and it is comparatively common in zoos; there are perhaps 300 monkeys in colonies in the United States and Europe. Although this trade contributed to the decline of the species in the wild, it is perhaps fortunate that there is a good breeding stock in captivity. Some of these captive tamarins have now been taken back to Brazil and reintroduced to the Poço das Antas Reserve. Although there have been many problems trying to teach ex-captive individuals to live in the wild, the project now appears to be succeeding. The juveniles have adapted comparatively well, and some of the adults have even begun to breed.

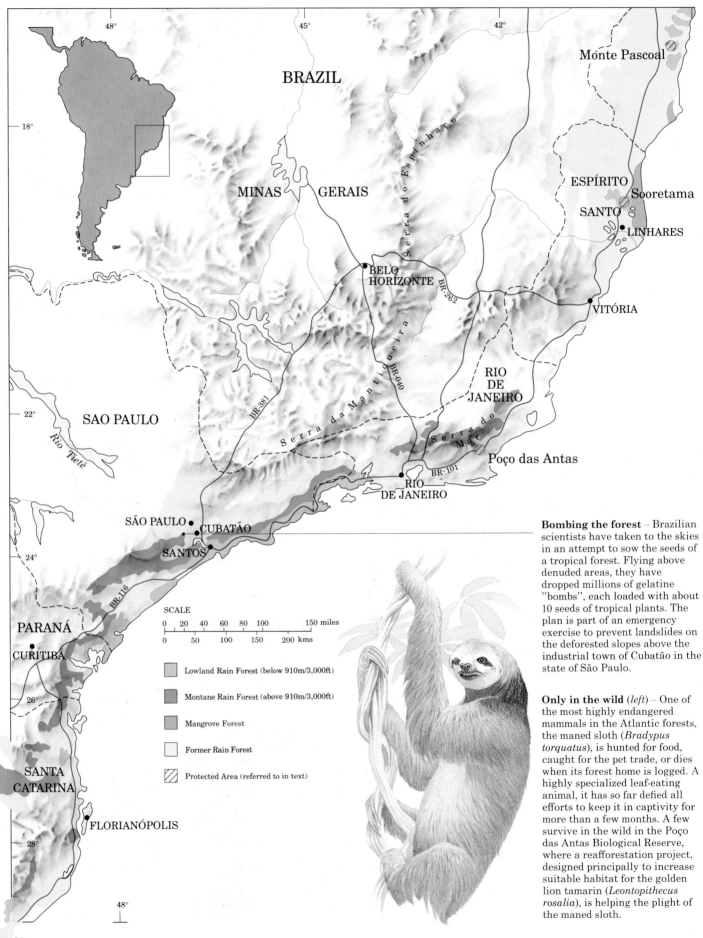

BRAZIL

MINAS   GERAIS

Serra do Espinhaço

ESPÍRITO
SANTO

Sooretama

LINHARES

BELO
HORIZONTE

BR-262

VITÓRIA

Serra da Mantiqueira

BR-040

RIO
DE
JANEIRO

SAO PAULO

BR-381

Serra do
Mar

Poço das Antas

Rio Tietê

BR-101

RIO
DE JANEIRO

SÃO PAULO

CUBATÃO

SANTOS

BR-116

PARANÁ

CURITIBA

SCALE

0   20 40 60 80 100   150 miles

0   50   100   150   200 kms

Lowland Rain Forest (below 910m/3,000ft)

Montane Rain Forest (above 910m/3,000ft)

Mangrove Forest

Former Rain Forest

Protected Area (referred to in text)

SANTA
CATARINA

FLORIANÓPOLIS

Monte Pascoal

Bombing the forest – Brazilian scientists have taken to the skies in an attempt to sow the seeds of a tropical forest. Flying above denuded areas, they have dropped millions of gelatine "bombs", each loaded with about 10 seeds of tropical plants. The plan is part of an emergency exercise to prevent landslides on the deforested slopes above the industrial town of Cubatão in the state of São Paulo.

Only in the wild (*left*) – One of the most highly endangered mammals in the Atlantic forests, the maned sloth (*Bradypus torquatus*), is hunted for food, caught for the pet trade, or dies when its forest home is logged. A highly specialized leaf-eating animal, it has so far defied all efforts to keep it in captivity for more than a few months. A few survive in the wild in the Poço das Antas Biological Reserve, where a reafforestation project, designed principally to increase suitable habitat for the golden lion tamarin (*Leontopithecus rosalia*), is helping the plight of the maned sloth.

The forests were first exploited for their timber, and then the fertile lands of the coastal plains were converted to agricultural plantations, particularly for the cultivation of sugar cane. With the discovery of gold and diamonds in the late sixteenth century, a move inland started and the wilderness areas of Minas Gerais and São Paulo became heavily populated, whole forests being cleared for the mines and farms needed to feed the miners. The mines were exhausted within a century. Agriculture then became the most important economic activity in Brazil, which meant that vast areas of forest were cleared for crops such as coffee, bananas and rubber. Even the wars with the Indians took their toll of the forests, for military divisions used to set fire to areas to drive the natives out of their homes and hiding places.

As early as 1831, the French naturalist Auguste de Saint-Hilaire expressed concern for the fate of the magnificent forests that were steadily being changed to farmland. However, from the beginning of this century, and particularly in the last two or three decades, the destruction has accelerated.

The Atlantic forest region is now the agricultural and industrial heart of Brazil. It has within its borders two of the three largest cities in South America, Rio de Janeiro and São Paulo (the latter is one of the largest in the world). Forty-three percent of Brazil's rapidly growing population of around 148 million people is squeezed into this region. The forests are now tiny green islands in a sea of civilization.

Virtually all of the Atlantic forest region is in private hands. Although under current laws 20 percent of this has to be kept as forest, the fines for violating these regulations are only a fraction of the income that can be obtained by selling the wood. This has created a mosaic of forest fragments, which display a remarkable biological robustness in that there are no documented extinctions in this region. Sustainable forest management has yet to be attempted. Frequently, the techniques used for the extraction of a few timber species causes widespread destruction.

At the moment a mere 0.1 percent of the original forest expanse is protected in national parks, biological reserves, ecological stations, state parks and private reserves. The conservation of these areas is hampered by lack of finances and inadequate management, and there are no means of enforcing existing protective legislation. There is now little or no suitable land available for the establishment of new reserves and parks, except perhaps in southern Bahia, so it is essential that the existing areas are adequately protected.

Fortunately, rural land owners are interested in creating their own private fauna refuges and the programme to encourage this has been very successful. There is also every sign that the younger generation in Brazil is interested in and concerned about the environment. There has been much pressure from both outside and within the country to preserve the last fragments of the Atlantic forest.

Saved from railway – A mining company, cvrd (Companhia Vale do Rio Doce), owns a remarkably well preserved stretch of rain forest next to the Sooretama National Biological Reserve near Linhares in the state of Espírito Santo. The company originally acquired the site to cut down the forest and use the wood for railway sleepers. However, the decline of mining in the area made the railway unnecessary and this, combined with pressure from the World Bank and, most significantly, its own realization of the value of the forest as it stood, decided the area's fate. Instead of being destroyed, the forest is now used as a research site by the company. The supervisor of the area, Renato de Jesus, is a botanist and he has found many new plant species and even a completely new plant family. Native plants from the reserve are used to restore areas affected by the company's mining activities elsewhere, and there are teams of seed gatherers in the forest each day. In spite of the high cost of the reserve, about US$350,000 a year, the company is committed to study and protect the area. This is, along with the Sooretama Reserve, the largest remaining relatively intact stretch of lowland forest in the entire Atlantic forest region.

Threatened birds – Although all hunting of wildlife is prohibited in the Atlantic forests, law enforcement outside the reserves and national parks is almost non-existent. Larger bird species, including the red-ruffed fruitcrow (*Pyroderus scutatus*), the dusky-legged guan (*Penelope obscura*) and the rusty-margined guan (*Penelope superciliaris*) pictured above, have been almost hunted out in some areas; but even tiny manakins and flycatchers are considered fair game. Another serious threat to the birds is the massive cage-bird trade. In rural areas, it is uncommon to find a house without one or more birds in cages, and most town and city houses have them too. Seed eaters are particularly popular; macaws and parrots are also in demand. All of these birds are now hard to find in the wild. It is possible that some of the rarer species would be able to spread out from the patches of primary forest in which they survive at present if the surrounding forest was protected, acting as a buffer zone. Some areas have been reafforested with exotic tree species such as eucalyptus and pine, but this does not help the birds because most of them cannot survive in these forests. Their numbers are also reduced by the use of pesticides on farms near to the primary forest.

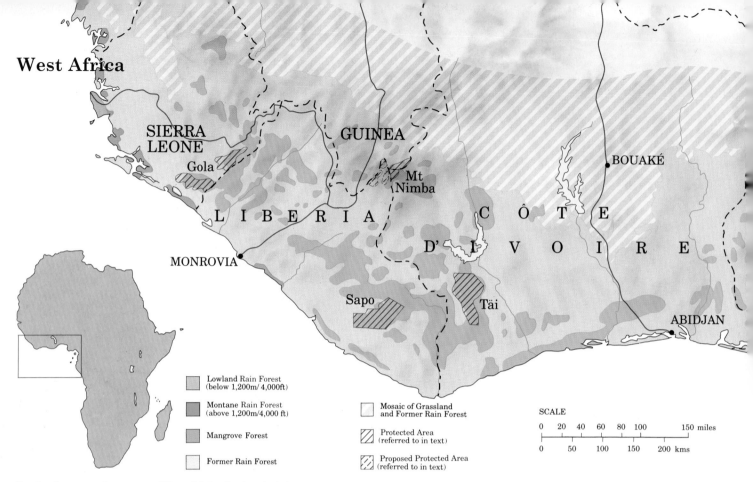

West Africa

Lowland Rain Forest
(below 1,200m/ 4,000ft)

Montane Rain Forest
(above 1,200m/4,000 ft)

Mangrove Forest

Former Rain Forest

Mosaic of Grassland
and Former Rain Forest

Protected Area
(referred to in text)

Proposed Protected Area
(referred to in text)

SCALE

0   20   40   60   80   100   150 miles

0   50   100   150   200 kms

By the fourteenth century, West Africa had a thriving economy and social structure based on trading gold across the Sahara. Following the arrival of Europeans, a lucrative coastal trade in gold, ivory and slaves was also established. The infamous slave trade was not stopped until the latter half of the eighteenth century, just prior to Europe's scramble for dominion over Africa. Cash crops such as oil palm, cocoa, ground nuts and cotton then became important throughout the region, and staple foods included cassava (introduced from South America, where it is called manioc), yam and maize (introduced from Central America). By the 1960s many West African nations had gained independence, and minerals such as oil, uranium, iron ore, gold and diamonds had become important economically.

Originally, the West African belt of moist tropical forest was virtually uninhabited, because the population favoured the more comfortable savanna farther inland. The introduction of commercial logging, the advent of modern tools and increased population pressure has changed all that.

The West African evergreen forests contain many economically important plants including the African rubber tree (*Landolphia* sp.), the Sierra Leone frankincense (*Daniellia thurifera*), yam (*Dioscorea rotunda*), oil palm (*Elaeis guineensis*) and kapok (*Ceiba pentandra*). They are also home to the African elephant (*Loxodonta africana*), pygmy hippo (*Choeropsis liberiensis*), pygmy chimpanzee (*Pan paniscus*), drill (*Papio leucophaeus*), red colobus monkey (*Colobus badius*) and many other rarities. There are quite a number of species unique to the region: in the Tai forest in Côte d'Ivoire, 54 percent of the 1,300 plant species are endemic. There is also a rich bird and butterfly fauna.

The population of West Africa is growing at a rate close to the human maximum – doubling every 20 years. The largest country by far is Nigeria, home for one in every five Africans. Immigration from the drought-stricken Sahel further swells numbers in some countries, such as Mali and Niger. The population problem is compounded by the fact that many West African governments are economically dependent on cash crops, which take up land at the expense of food crops, and by the relative lack of applied research into ways of improving local agriculture. The genetic engineering that resulted in the hybrid cereals used in the Asian and Latin American Green Revolutions largely bypassed Africa. Low soil

fertility and poor agricultural techniques continue to produce low yields, which in turn lead to repeated clearance of rain forest for agriculture, with the ensuing damage to watersheds, soil structure and water supplies. The burgeoning population also has an increasing need for construction materials and fuelwood. Traditional cooking methods often waste up to 90 percent of the heat available from burning wood, when relatively simple changes could halve this loss.

Commercial logging started in the more accessible regions in the 1880s involving the selective extraction of African mahogany (*Khaya ivorensis*) and limbe (*Terminalia superba*). At that time there was about 420,000 square kilometres (160,000 square miles) of rain forest. One hundred years later, in 1980, only 173,000 square kilometres (66,800 square miles) remained, and a further four percent was being lost each year. Today, perhaps 166,000

The population of West Africa is growing at a rate close to the human maximum – doubling every 20 years. The largest country...is Nigeria, home for one in every five Africans.

square kilometres (64,000 square miles) is still standing. The primary forests of Sierra Leone, Guinea, Côte d'Ivoire and Nigeria have all decreased to less than 10 percent of their original extent, although Ghana retains a more healthy 22 percent.

Recent developments give cause for some optimism. Liberia, Sierra Leone and Nigeria are at last gazetting rain forest national parks. The conservation movement has come a long way from its original stance of maintaining sacrosanct reserves. The new approach emphasizes dialogue with rural peoples to protect core areas, surrounded by buffer zones where the sustainable use of the land is encouraged.

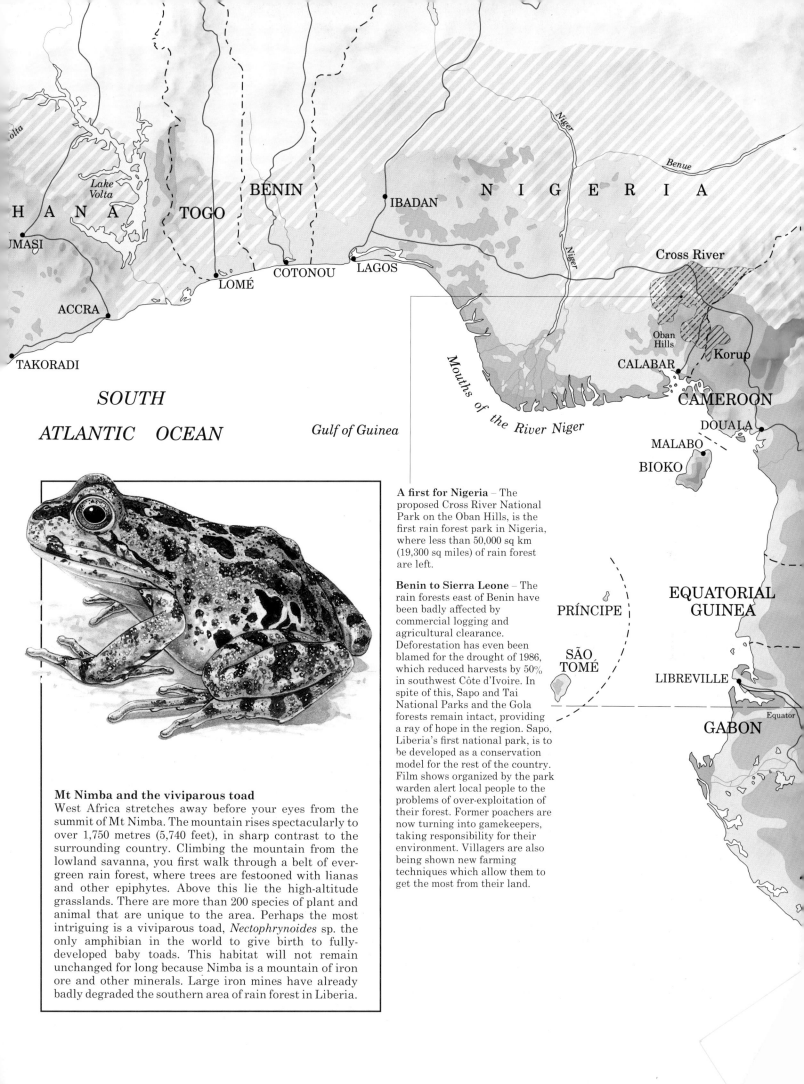

SOUTH

ATLANTIC OCEAN

Gulf of Guinea

HANA
UMASI

TOGO

BENIN

LOMÉ
COTONOU
LAGOS
ACCRA
TAKORADI

IBADAN

Lake Volta

Volta

N I G E R I A

Niger

Benue

Niger

Cross River

Oban Hills

CALABAR

Korup

CAMEROON

DOUALA

MALABO

BIOKO

Mouths of the River Niger

EQUATORIAL GUINEA

PRÍNCIPE

SÃO TOMÉ

LIBREVILLE

Equator

GABON

A first for Nigeria – The proposed Cross River National Park on the Oban Hills, is the first rain forest park in Nigeria, where less than 50,000 sq km (19,300 sq miles) of rain forest are left.

Benin to Sierra Leone – The rain forests east of Benin have been badly affected by commercial logging and agricultural clearance. Deforestation has even been blamed for the drought of 1986, which reduced harvests by 50% in southwest Côte d'Ivoire. In spite of this, Sapo and Tai National Parks and the Gola forests remain intact, providing a ray of hope in the region. Sapo, Liberia's first national park, is to be developed as a conservation model for the rest of the country. Film shows organized by the park warden alert local people to the problems of over-exploitation of their forest. Former poachers are now turning into gamekeepers, taking responsibility for their environment. Villagers are also being shown new farming techniques which allow them to get the most from their land.

Mt Nimba and the viviparous toad

West Africa stretches away before your eyes from the summit of Mt Nimba. The mountain rises spectacularly to over 1,750 metres (5,740 feet), in sharp contrast to the surrounding country. Climbing the mountain from the lowland savanna, you first walk through a belt of evergreen rain forest, where trees are festooned with lianas and other epiphytes. Above this lie the high-altitude grasslands. There are more than 200 species of plant and animal that are unique to the area. Perhaps the most intriguing is a viviparous toad, *Nectophrynoides* sp. the only amphibian in the world to give birth to fully-developed baby toads. This habitat will not remain unchanged for long because Nimba is a mountain of iron ore and other minerals. Large iron mines have already badly degraded the southern area of rain forest in Liberia.

African grey parrot

Senegal parrot

Trading in parrots – West Africa has a wonderful variety of parrots from the green-yellow Senegal (*Poicephalus senegalus*) to the talking African grey (*Psittacus erithacus*). It is also the largest bird-exporting region in the world, more than 75% of the captured birds being sent to the United States and Europe as pets. For every bird that completes its journey, 10 die during capture or transit. The British market could now be solely supplied by existing captive-breeding projects, but such birds are twice as expensive as those caught in the wild. Encouragingly, New York State banned imports of caged wild birds in 1984, and there have even been proposals for villagers to breed the African grey parrot in the buffer zone around Salonga National Park in Zaire.

With local consent – Villagers in Cameroon and Nigeria often gather to discuss important local matters, or to celebrate a festival (*left*), but now there is a new reason for gathering – the conservation of their local environment. Local meetings are the mainstay of the Oban Hills Project in Nigeria and the Korup Project in neighbouring Cameroon. They are aimed at gaining local consent for the conservation measures needed to protect the rain forests. Six villages have already agreed to be relocated outside the Korup Park in more fertile areas. Local meetings have also revealed that although most villagers hunt within the boundaries of the parks, the younger generation regards it as exhausting and dangerous work; so alternative employment, including the collection of forest honey and medicinal barks, is being encouraged.

The hunting of mermaids – Manatees, or sea cows, are the ancient source of mariner's tales of mermaids. Unlike other marine mammals, they are herbivorous and, as well as living along coastlines, they can move inshore to coastal swamps and even up rivers. Manatees have a low reproductive rate and the mother suckles her young for about 18 months. They are particularly docile and are hunted for their delicious meat. Regarded as pests by fishermen, the Senegal manatee (*Trichechus senegalensis*) is caught by trapping, netting and harpooning. The hunters rely on the fact that the manatees move into mangrove swamps as the tide rises; to catch them, nets and traps are set to prevent their escape as the tide ebbs. They are worth little in cash terms because the meat is customarily shared among the villagers. In a survey in 1987, some manatee trappers in Sierra Leone reported a decreased catch, but other areas still support large populations. There is considerable potential for manatee conservation.

Rediscovered gorillas – Western lowland gorillas (*Gorilla gorilla gorilla*) have been rediscovered in Nigeria after 30 years during which they were believed to have become extinct there. This find is all the more astonishing considering the density of Nigeria's human population. Gorillas are threatened by habitat loss and hunting for food and trophies. A single gorilla carcass can fetch double a labourer's monthly wage in Lagos, and one community reported killing 6 gorillas in 1987 alone, not to mention those wounded. There may be up to 300 individuals in 3–5 groups within the country. Although highly elusive in thick forest, their numbers can be assessed from counting sleeping nests and their distinctive, trilobed droppings. The studies have shown that the gorillas are still reproducing, but the births are not keeping pace with the numbers being killed through hunting. Nigeria is now considering the creation of a new park, the Boshe-Okawango National Park, to protect the western lowland gorilla.

Food from the wild
Up to one-fifth of all the animal protein consumed in rural areas of West Africa comes from wild animals. This "bushmeat" includes small antelopes, such as duikers, and a wide variety of monkeys; less obvious culinary delicacies include cane rats, porcupine, caterpillars, termites, grasshoppers and the giant African land snail (*Achatina fulica*) pictured above. In much of the region, bushmeat is preferred to domestic livestock and is more expensive. Continued hunting pressure is threatening a number of endangered species, particularly primates. Wildlife farming, coupled with revision of existing legislation to allow controlled hunting of non-threatened animals, may ensure the survival of rare primates and antelopes, yet allow bushmeat still to be eaten.

Central and East Africa

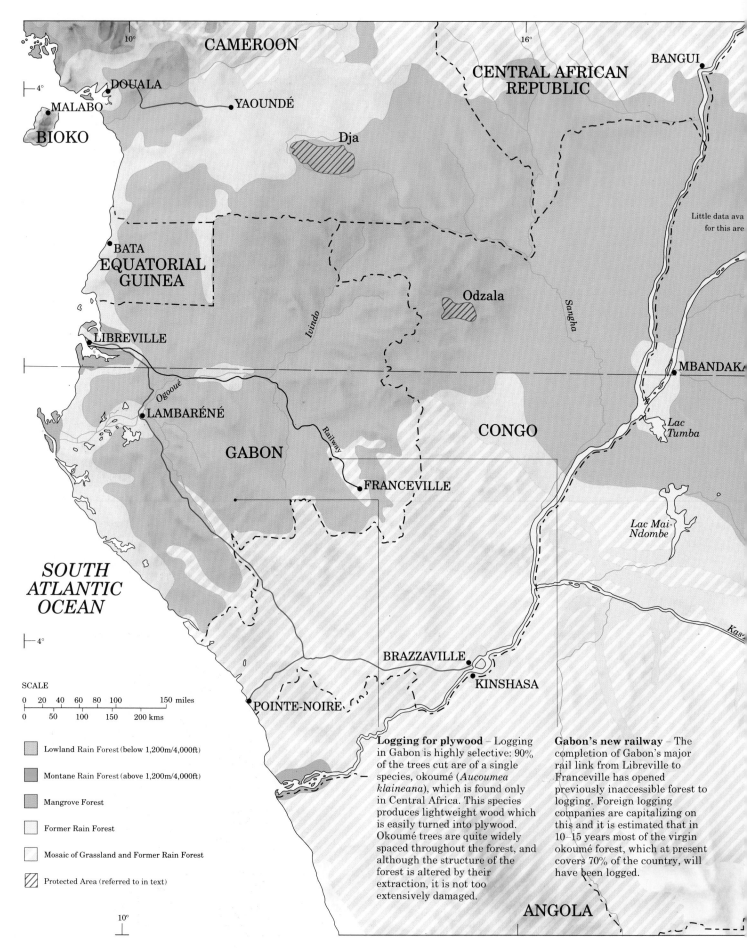

SCALE

| 0 | 20 | 40 | 60 | 80 | 100 | | 150 miles |
| 0 | 50 | 100 | 150 | | 200 kms | | |

Lowland Rain Forest (below 1,200m/4,000ft)

Montane Rain Forest (above 1,200m/4,000ft)

Mangrove Forest

Former Rain Forest

Mosaic of Grassland and Former Rain Forest

Protected Area (referred to in text)

Logging for plywood – Logging in Gabon is highly selective: 90% of the trees cut are of a single species, okoumé (*Aucoumea klaineana*), which is found only in Central Africa. This species produces lightweight wood which is easily turned into plywood. Okoumé trees are quite widely spaced throughout the forest, and although the structure of the forest is altered by their extraction, it is not too extensively damaged.

Gabon's new railway – The completion of Gabon's major rail link from Libreville to Franceville has opened previously inaccessible forest to logging. Foreign logging companies are capitalizing on this and it is estimated that in 10–15 years most of the virgin okoumé forest, which at present covers 70% of the country, will have been logged.

22°

28°

Ubanji

4°

Lake Albert

*re
ngo)*

KISANGANI

Equator

*Lake
Edward* UGANDA

Z A I R E

RWANDA

*Lake
Kivu* KIGALI

Salonga

Kilbira

Ruvubu
BUJUMBURA

BURUNDI

4°

Lake Tanganyika

TANZANIA

Mbuti – The Ituri forest, home of the hunter-gathering Mbuti Pygmies, is threatened by human settlement from outside. These audacious Pygmies fell forest elephants by cutting their hamstrings, but also catch many other animals including the blue duiker (*Cephalophus monticola*) and the sitatunga (*Tragelaphus spekei*). Fortunately, they can find most of their traditional prey in the areas of regenerating secondary forest that are left after the settlers have moved on.

22°

28°

0

8°

A belt of tropical rain forest spans the centre of Africa, running from Cameroon and Gabon on the Atlantic coast to Kenya and Tanzania on the Indian Ocean. Within this belt, the climate, and hence the type of forest, is very varied. Unending vistas of dark, impenetrable jungle are associated with the Central African countries, while East Africa is largely covered with bushland, the rain forests restricted to fertile mountain regions.

More than 80 percent of Africa's rain forest is in Central Africa, much of it largely untouched. Relatively little is known about the region's history, and there are remarkably few excavated archaeological sites. The vast Zaire Basin is inhospitable to settlers. In contrast, the East African climate is more favourable to human settlement and agriculture, and some of the earliest human remains have been discovered there. Agriculture is the mainstay of most East African economies, superseded by tourism in Kenya, whereas the Central African economies are largely based on minerals – diamonds in the Central African Republic, copper and cobalt in Zaire, manganese and oil in Gabon.

The rain forests of Central Africa are believed to originate from a number of small, isolated patches of forest, called refugia (see page 52), that survived the dry African climate during the last Ice Age. These refugia, one of which was in the highlands of eastern Zaire, are extremely old and support a wealth of species. When the climate became wetter, 12,000 years ago, the rain forest expanded, recolonizing the region. Today, Zaire is Africa's richest country in terms of plants and animals, having more than 11,000 plant species of which nearly one-third are found nowhere else. With a grand total of 409 species of mammal, it has almost 100 more than any other country on the continent. The forests of Central Africa are home to such animals as pygmy chimpanzees, gorillas, okapis, and forest elephants. The discovery of the suntailed guenon (*Cercopithecus solatus*), a new species of primate, in Gabon, and the rediscovery of the eastern lowland gorilla (*Gorilla gorilla graueri*) in western Zaire, suggest that

Tanzania's mountain forests
The eastern rain forests of Tanzania, particularly those on the Usambara, Uluguru and Uzungwa mountains, may have received a reprieve. These ancient mountains, islands of heavy rainfall in an otherwise arid country, were uninhabited until Europeans set up coffee and tea plantations at the beginning of this century. But the population increase in recent decades, and the development of timber extraction, have threatened the mountains' unique ecology. Recently a plan to set up commercial sawmills has been replaced by an IUCN project aimed at watershed conservation and sustainable development. Poor farming techniques have led to soil erosion, and the need to clear more forest has been increasing. To combat this problem, new crops and farming techniques are being encouraged in the eastern Usambaras. Contour planting and nitrogen-fixing shade trees have been introduced, so that the farmers are able to protect their land, and villagers are managing areas of forest attributed to them.

Elephants and conservation – The international ivory trade has halved elephant numbers in Africa during the last decade. In the Zaire Basin elephants continue to survive well away from human settlement, but the relationship between elephants, Pygmies and forest regeneration that has existed for centuries is in jeopardy. In their quest for salt, elephants dig at the ground (*below*), opening up small patches of the forest, allowing light to penetrate which encourages the growth of understorey plants. This new vegetation attracts antelope which are then hunted by Pygmies. In January 1990, a worldwide ban on the ivory trade came into force. However, several African nations have said they will maintain the trade while protecting their herds.

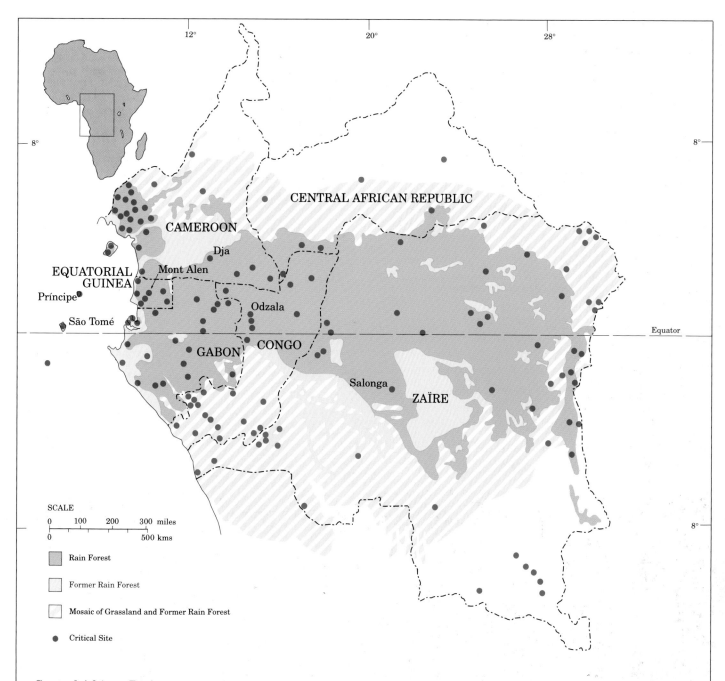

SCALE

| 0 | 100 | 200 | 300 miles |

| 0 | 500 kms |

Rain Forest

Former Rain Forest

Mosaic of Grassland and Former Rain Forest

● Critical Site

Central African Project

An ambitious project for the conservation and sustained use of forest ecosystems, designed by IUCN and funded by the EEC, is under way in Central Africa. More than 100 sites of critical importance to the safeguarding of biodiversity have been identified and measures proposed to ensure their conservation. A network of pilot projects is being set up, one in each of the countries of the region, in which different aspects of a community-based approach to resource management will be demonstrated.

In the Central African Republic, post-logging forest management will be examined as a way of enriching the forest to encourage investment in its maintenance.

In Cameroon, genetically improved hardwood species will be bred and high-yielding plantations established in the buffer zone around the Dja Reserve.

In Gabon, the aim is to improve the sustainability of

traditional hunting by giving villagers control over local hunting areas.

In São Tomé, plantations of fast-growing species will provide fuelwood and construction timber to relieve demand for these products from natural forests.

In Congo, application of the Pygmies' traditional knowledge of the forest ecosystem will be used in the buffer zone around the Odzala National Park.

In Equatorial Guinea, forest concessions will be zoned to provide protection to the Mont Alen National Park, and agroforestry will be encouraged in communities living near to the edge of the forests.

In central Zaïre, a major research centre will be established, near to the Salonga National Park, as a regional focus for research into the many different ways in which the forest can be used and conserved.

further equally exciting finds are still to be made in this region.

Hunter-gathering, one of the oldest ways in which humans have exploited their environment, has long been practised in the rain forests of Central Africa. Even today the Babinga Pygmies of the Central African Republic and the Mbuti of Zaire and Congo continue to harvest the forest in this way. In the Ituri forest, the Mbuti have been closely associated with a tribe of shifting cultivators, the Bantus, for more than 2,000 years. The Bantus exchange the products of their agriculture for the Mbuti's fruits of the forest. The way in which the Bantus carry on this ancient system of agriculture, in which patches of forest are cleared for a few years of subsistence cropping, does little long-term damage. During long fallow periods the forest regenerates around the cut tree trunks which have been left behind. Increasing population pressure and settlement by urban poor is disrupting many of the traditional agricultural techniques, however, leading to shorter fallow periods and over-working of the fragile soils.

As the population increases, the need for development is affecting more of the region. Roads have been built through Zaire and the Central African Republic. The trans-Gabon railway was completed in 1987, encouraging further settlement and forest encroachment. Zaire has a population density of only 13 people per square kilometre (34 people per square mile), 25 times lower than that of neighbouring Rwanda. But the rate of growth of its

---

The rain forests of Central Africa are believed to originate from a number of small, isolated patches of forest...that survived the dry African climate during the last Ice Age.

---

population has been increasing over the last 30 years to the present rate of more than three percent a year. In two or three decades Zaire's population will have doubled, and there will be even more pressure to colonize the forests.

Surrounded by people – The mountain forests of Rwanda and Burundi, home of the mountain gorilla (*Gorilla gorilla beringei*), are surrounded by a sea of humanity. These tiny countries support the continent's highest population densities, having a staggering 350 people per sq km (906 people per sq mile). The process of forest clearance, *kwica ishyamba*, which literally means "to kill the forest", was started by the Hutu agriculturalists on payment of dues to the original custodians of the forests, the hunter-gatherer Twa tribe. In the 13th century, the pastoralist Tutsi invaded and since then pastoralism has been the dominant system of agriculture. Recent forest clearance has been for cash crops such as pyrethrum, tea, coffee and quinine. Encouragingly, this is now on the decline, and Burundi has established its first 2 national parks, Kilbira and Ruvubu, within the past 10 years. Rwanda is already well known for its highly successful mountain gorilla project.

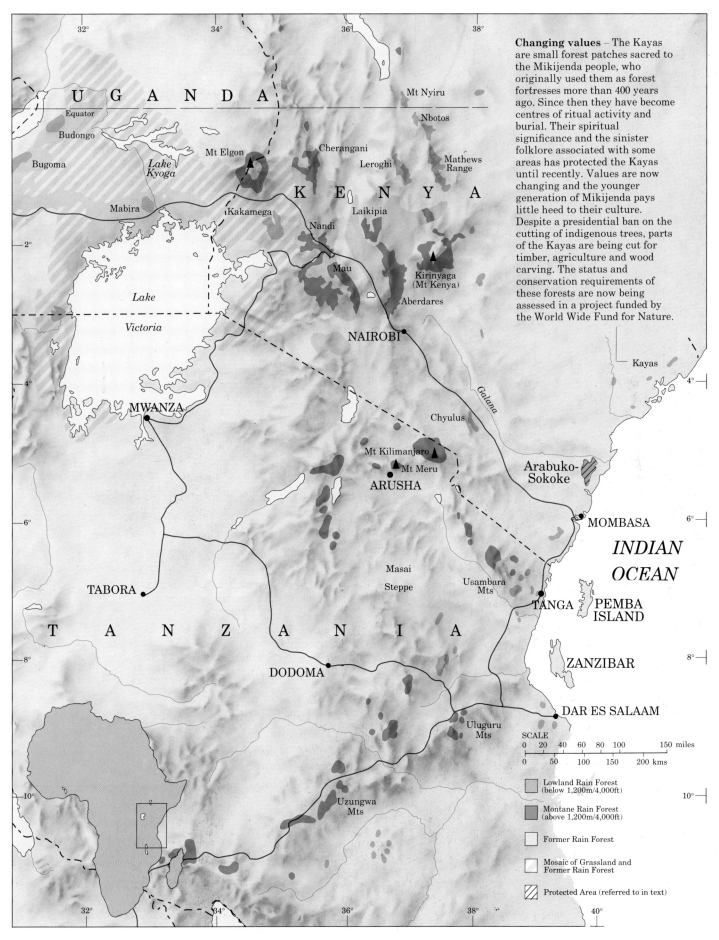

Changing values – The Kayas are small forest patches sacred to the Mikijenda people, who originally used them as forest fortresses more than 400 years ago. Since then they have become centres of ritual activity and burial. Their spiritual significance and the sinister folklore associated with some areas has protected the Kayas until recently. Values are now changing and the younger generation of Mikijenda pays little heed to their culture. Despite a presidential ban on the cutting of indigenous trees, parts of the Kayas are being cut for timber, agriculture and wood carving. The status and conservation requirements of these forests are now being assessed in a project funded by the World Wide Fund for Nature.

SCALE

| 0 | 20 | 40 | 60 | 80 | 100 | | 150 miles |
| 0 | 50 | | 100 | 150 | | 200 kms |

Lowland Rain Forest (below 1,200m/4,000ft)

Montane Rain Forest (above 1,200m/4,000ft)

Former Rain Forest

Mosaic of Grassland and Former Rain Forest

Protected Area (referred to in text)

145

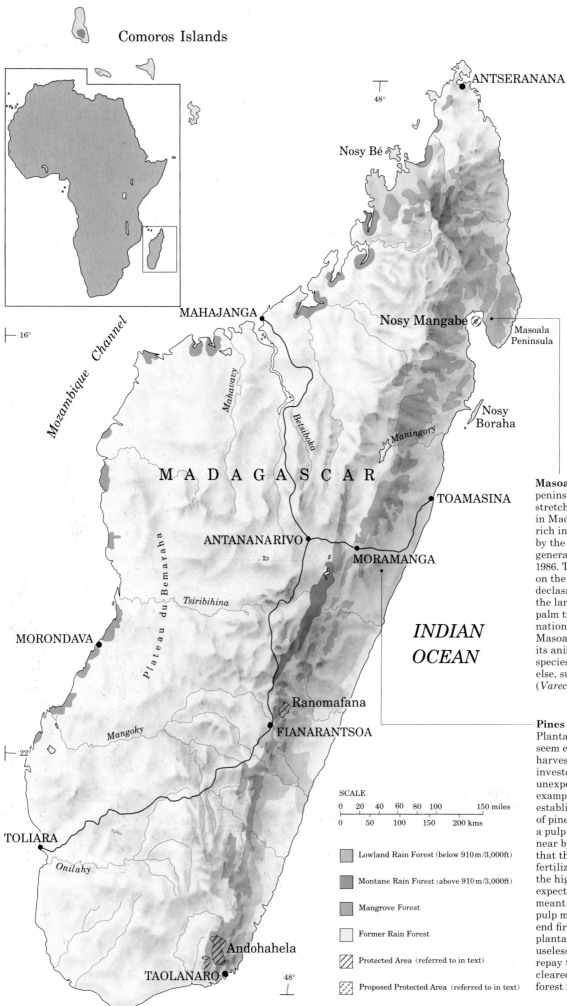

Comoros Islands

ANTSERANANA

48°

Nosy Bé

16°

Mozambique Channel

MAHAJANGA

Nosy Mangabe

Masoala
Peninsula

MADAGASCAR

Mahavay

Betsiboka

Maningory

Nosy
Boraha

TOAMASINA

ANTANANARIVO

MORAMANGA

Plateau du Bemaraha

Tsiribihina

INDIAN
OCEAN

MORONDAVA

Mangoky

Ranomafana

22°

FIANARANTSOA

TOLIARA

Onilahy

SCALE

0   20   40   60   80   100   150 miles

0   50   100   150   200 kms

Andohahela

TAOLANARO

48°

Lowland Rain Forest (below 910 m/3,000 ft)

Montane Rain Forest (above 910 m/3,000 ft)

Mangrove Forest

Former Rain Forest

Protected Area (referred to in text)

Proposed Protected Area (referred to in text)

Masoala – The Masoala
peninsula contains the largest
stretch of lowland rain forest left
in Madagascar. It is particularly
rich in endemic plants, as shown
by the fact that 2 new palm
genera were discovered there in
1986. There used to be a reserve
on the peninsula, but it was
declassified in 1964 and some of
the land is now planted with oil
palm trees. It is hoped that a
national park will be set up on
Masoala to protect the forest and
its animals, particularly those
species that are found nowhere
else, such as the red-ruffed lemur
(*Varecia variegata rubra*).

Pines that didn't pay –
Plantations of exotic trees often
seem easier to manage and
harvest than natural forests, but
investors have run into
unexpected difficulties. For
example, the World Bank helped
establish 600 sq km (230 sq miles)
of pines near Moramanga to feed
a pulp mill that was to be built
near by. It was soon discovered
that the trees needed expensive
fertilizers if they were to reach
the high levels of productivity
expected of them. Poor growth
meant that plans for the new
pulp mill were abandoned. In the
end fire destroyed some of the
plantation, and the rest is
useless. Madagascar now has to
repay the loan for a project that
cleared potentially useful natural
forest for little or no benefit.

At around 1,600 kilometres (1,000 miles) long and 450 kilometres (280 miles) wide, Madagascar is the world's fourth largest island. It has been separated from the mainland for 150 million years or more. Most of its plants and animals have evolved in isolation for at least 40 million years, ever since the channel between Africa and Madagascar became too wide for any life to raft across from the mainland. The result is that most of Madagascar's flora and fauna is unique to the island. Botanically, it is one of the richest areas in the world.

About 80 percent of Madagascar's 10,000 plant species are endemic. All 30 of its primate species are lemurs, and none is found anywhere outside Madagascar and the adjacent Comoro Islands. Two-thirds of the world's chameleons – from one the size of a thumbnail to one 60 centimetres (24 inches) long – are found on the island. In addition, more than 90 percent of Madagascar's reptiles and amphibians are unique to the island. The list could continue. However, so much of this wildlife is threatened that Madagascar is frequently considered to be the single highest conservation priority in the world.

Rural population pressure

Humans arrived on Madagascar as recently as 1,500 years ago, many of them having come originally from Indonesia. Today, there are 11.2 million people in the country, more than twice as many as there were in 1960, and it is estimated there will be 28 million by the year 2025. The population is still mostly rural and dependent on the land for its livelihood. It is this rural population pressure that is causing the rapid destruction of the rain forests that lie along the eastern side of the country.

To feed the expanding population, the forests are cut and burned and rice is most commonly planted in place of the trees; but maize, cassava (manioc) and other crops are also grown. As with most other rain forest areas, the soil is relatively impoverished and sustains the crops for only a couple of years. Falling yields force the farmers to move on and clear more forest. The forests also supply the rural population with all its fuel. Because almost all the lowland forest has disappeared, the search for fuelwood is now reaching higher up the hillsides. In 1985, it was estimated that as little as 38,000 square kilometres (14,700 square miles) of rain forest remained intact, half that recorded in 1950 and about one-third of the original extent. If cutting continues at the present rate, only those forests in the most remote and inaccessible areas will survive the next 35 years.

Destruction of the Malagasy rain forest has been described as "a tragedy without villains". Although there has been some commercial logging in the past, particularly for ebony and rosewood, this has had comparatively little impact. The major threat is from the small-scale farmers and their slash-and-burn, or *tavy*, agriculture. The government is committed to incorporating conservation into the country's development strategy, and the international community is providing money and scientific expertise. Without this there is little hope that Madagascar will continue to live up to the description given it by an eighteenth-century explorer as "the naturalist's Promised Land".

Wallace's moth – Madagascar has more than 1,000 species of orchid. One of these, the comet orchid (*Angraecum sesquipedale*), has a tubular nectary that is 35 cm (14 in) long. When Alfred Russel Wallace, an eminent Victorian botanist and contemporary of Charles Darwin, found this plant he predicted that it must be fertilized by a moth with a tongue at least 35 cm in length, so that the moth could reach the nectar. In due course the moth (*Xanthopan morgani*) was found and given the subspecific name *praedicta*, in honour of Wallace's foresight.

Cyanide bamboo – Discovered in 1987, the golden bamboo lemur (*Hapalemur aureus*) is one of the rarest primates in Madagascar. It also has one of the strangest diets, for it feeds on a plant that might be expected to kill it. The giant bamboo (*Cephalostachyum vigueri*), pictured above, contains cyanide, and the amount that the lemur eats each day would be fatal to a human.

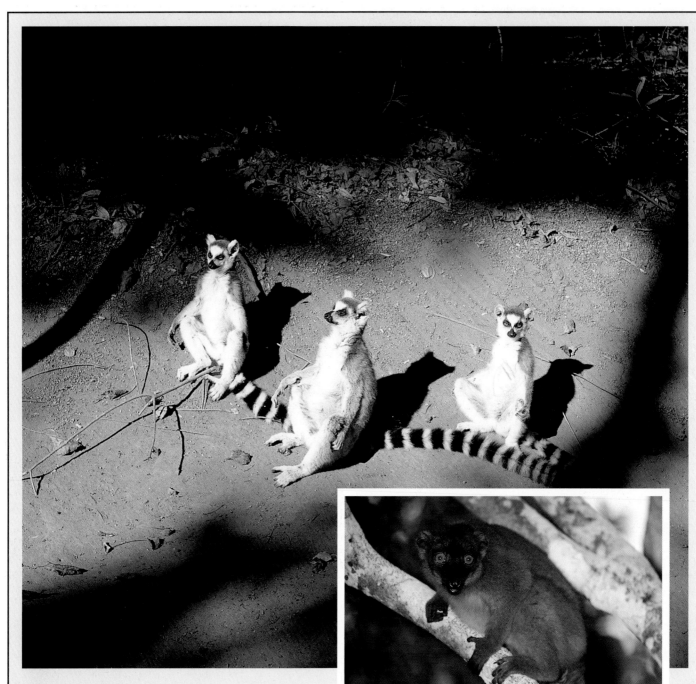

(*Main pic*) Sunbathing ring-tailed lemurs (*Lemur catta*)
(*Inset*) Mayotte lemur (*Lemur fulvus mayottensis*)

Ghost primates

Probably best known of all Madagascar's animals are the lemurs. The name of these primates is derived from the Latin for ghosts – *lemures* – and if the Malagasy rain forests continue to be destroyed at the present rate, that is indeed what they may become.

There are 30 different species, ranging in size from the tiny, nocturnal ground mouse lemur (*Microcebus rufus*), which weighs a mere 60 grammes (two ounces), to the indri (*Indri indri*), a diurnal, family-living species which may weigh as much as ten kilogrammes (21 pounds). The eastern rain forests are home to more than half the lemur species, including the peculiar-looking aye-aye (*Daubentonia madagascariensis*), which is generally considered to be a harbinger of evil and may be killed on sight.

Before humans arrived in Madagascar, there were at least 14 more species of lemur on the island than there are today. Most were bigger than those alive now; the largest may have been about the size of a female gorilla. Their remains have been found alongside the pots that they were cooked in, and it appears that they became extinct through a combination of hunting and habitat destruction. These large species were the ones most vulnerable to the impact of humans, but the threats have not disappeared; although hunting may not be such a problem for the remaining, smaller lemurs, the destruction of the forests for agriculture continues at an accelerating pace. No forests means no lemurs, and the world will be a poorer place for their loss.

Male white-fronted lemur (*Fulvus albifrons*)

Ruffed lemur (*Varecia variegatus*)

Ranomafana

Integrating conservation, education, rural development and improvements in local living standards is the key to the Ranomafana National Park in southeastern Madagascar. From the beginning of the project in the mid-1980s, it was appreciated that the willing cooperation of the local people was vital; only if they become the forest managers, responsible for and reaping the benefits of the proposed changes, will the project succeed.

The plan is to have a core area of 416 square kilometres (161 square miles) to the park, surrounded by a 1,500 square kilometre (580 square mile) buffer zone. Within this zone the existing, destructive, *tavy* agricultural system will be improved and alternative farming methods will be introduced. These new methods may follow the example set in nearby Kianjavato, where model farms have been established to demonstrate alternative ways of using the land. These include reafforestation of upper watersheds, mixed cropping of middle slopes and intensive production in the bottom of valleys, with rice cultivation in paddy fields, fish farming and vegetable gardening.

There is also great potential for both international and national tourism within the proposed park. Already the area receives a small but increasing flow of visitors, and this can be stimulated to generate revenue for the maintenance of the park. There are few other places in Madagascar where, if you are lucky, you will see 12 different lemur species, let alone more than 70 different birds and numerous fascinating insects and plants.

With funding from agencies such as the World Wide Fund for Nature and USAID, and help from Duke and North Carolina State Universities, the Malagasy government is keen for Ranomafana National Park to become a reality.

Protecting the protected areas

As long ago as 1881 there was a law in Madagascar condemning those who cut down the forests to be put in chains. Then as now such laws were difficult to enforce. Madagascar was one of the first countries to set up a network of protected areas, which it did in 1927, but it does not have the resources needed to protect the reserves. Access to its 12 Integral Nature Reserves is strictly forbidden, except for scientific purposes. The Special Reserves, which are designed to protect particular plant or animal species, allow unlimited access, but hunting, fishing, grazing of livestock or removal of forest products is forbidden. Local villagers are allowed to exploit certain forest products in the country's two national parks.

These rain forest reserves vary in size from five square kilometres (two square miles) in the Special Reserve on the small island of Nosy Mangabe to 760 square kilometres (290 square miles) in the Andohahela Integral Nature Reserve. Many of them are protected by only one or two park guards, who generally do not live in the reserve itself but in nearby (or not so near) villages. They rarely have any transport, or the equipment needed for a patrol, and so it is impossible to ensure that the reserves are adequately safeguarded.

Since 1986, the World Wide Fund for Nature and the Malagasy Department of Water and Forests have been involved in a major project to survey all the protected areas in Madagascar. About many of the areas surprisingly little is known, not even what species of mammals and birds are present, let alone what plants and insects. Biological inventories have been compiled as part of the project, and sites of unusual diversity identified.

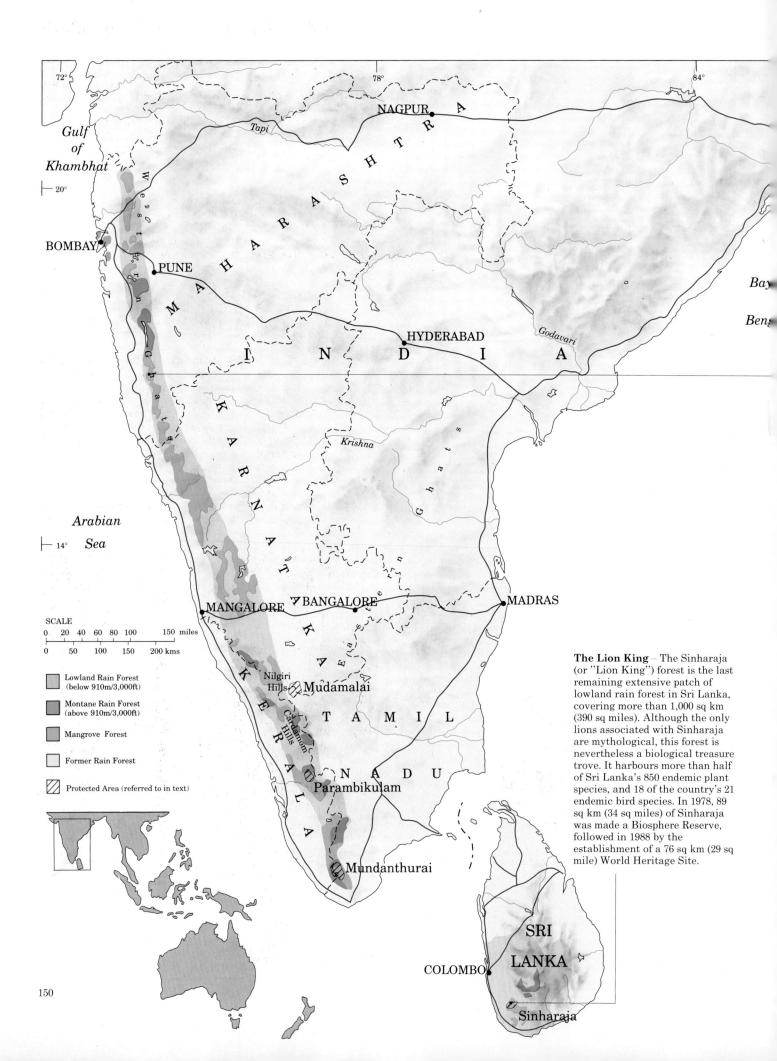

72° 78° 84°

NAGPUR

Gulf
of
Khambhat

Tapi

20°

M A H A R A S H T R A

BOMBAY

PUNE

Arabian

I N D I A

HYDERABAD

Godavari

14° *Sea*

K A R N A T A K A

Krishna

MANGALORE BANGALORE MADRAS

SCALE
0 20 40 60 80 100   150 miles
0   50   100   150   200 kms

Lowland Rain Forest
(below 910m/3,000ft)

Montane Rain Forest
(above 910m/3,000ft)

Mangrove Forest

Former Rain Forest

Protected Area (referred to in text)

Nilgiri
Hills Mudamalai

T A M I L

Cardamom
Hills

Parambikulam N A D U

K E R A L A

Mundanthurai

Bay

Beng

The Lion King – The Sinharaja
(or "Lion King") forest is the last
remaining extensive patch of
lowland rain forest in Sri Lanka,
covering more than 1,000 sq km
(390 sq miles). Although the only
lions associated with Sinharaja
are mythological, this forest is
nevertheless a biological treasure
trove. It harbours more than half
of Sri Lanka's 850 endemic plant
species, and 18 of the country's 21
endemic bird species. In 1978, 89
sq km (34 sq miles) of Sinharaja
was made a Biosphere Reserve,
followed in 1988 by the
establishment of a 76 sq km (29 sq
mile) World Heritage Site.

SRI

LANKA

COLOMBO

Sinharaja

India is home for about 15 percent of the world's people. Rain forests are found in the Western Ghats, Assam and northeast India, and the Andaman and Nicobar Islands. Each of these widely separated regions has its own endemic animals and plants, and many different peoples live in the forests. A few montane forests survive in Sri Lanka, and although Bangladesh has been almost completely deforested, the largest mangroves in the world are found in the Ganges delta.

India

Deforestation has been rapid in recent decades: up to 1,500 square kilometres (580 square miles) of forested land has been converted to other uses every year. More than 50,000 square kilometres (19,300 square miles) have been occupied by settlers or shifting cultivators, and the remaining forests often degraded by logging, fuelwood collection and clearance for grazing land (which is common in the Western Ghats).

India has about 45,000 plant species, as many as 4,500 of which are threatened with extinction. Medicinal plants are particularly at risk because they are widely collected by rural people to treat common ailments. The country has a tradition of establishing forest reserves and wildlife sanctuaries that goes back as far as the fourth century BC. In 1988, India had 66 national parks and 434 wildlife sanctuaries, but recent proposals aim to raise this to 148 and 503 respectively, covering in total area more than five percent of the country. Between 1950 and 1980 India lost large areas of her tropical forests, but in the 1980s a strong political will to protect and conserve forests developed, based on the force of public opinion. New policies were formulated and laws enacted to conserve what remains.

India: the Western Ghats

The last rain forests in peninsular India are found in the Western Ghats. These coastal hills extend from the extreme southern tip of India northwards to the Gulf of Khambhat. They cover 160,000

Sacred groves – The Indian word *shola* properly applies to rolling grassy hills with remnant patches of forest in sheltered sites beside rivers. Nowadays, it is used to describe any of the remaining rain forests of the Western Ghats. Within these *shola* forests there are a number of groves, each of which the locals believe to be under the protection of a particular god. All life within one of these groves is sacred, and so protected from any form of disturbance by the local people.

Trapped macaque – The rare lion-tailed macaque (*Macaca silenus*) occurs only in the evergreen forests of the Western Ghats, normally in groups of 15–20 animals. This primate can only travel through forest, so any gap in the canopy, due to a plantation or shifting cultivation, effectively blocks its path. Because of this, some populations are confined to islands of forest too small and isolated to allow sufficient genetic exchange with other troops. Wildlife managers plan to capture and exchange individuals to combat the problem of inbreeding.

Gaur on the Red List – The Mudumalai Wildlife Sanctuary in the Western Ghats has one of the largest gaur (*Bos gaurus*) populations in the world. With a huge head, massive body and sturdy limbs this animal is among the largest of all wild cattle. Once common throughout Asia, the gaur now survives only in isolated populations, and is listed as vulnerable to extinction in the IUCN Red List of Threatened Animals.

square kilometres (62,000 square miles), spanning the states of Maharashtra, Karnataka, Tamil Nadu and Kerala. About one-third of the Western Ghats is still covered by natural vegetation, of which 20,000 square kilometres (7,700 square miles) is rain forest. The rapidly expanding human population – now growing by 2.5 percent a year – has cleared or modified much of the region's original vegetation cover. Coastal lowland areas have been extensively settled, and the forests have made way for rice fields and coconut plantations. Tea, coffee, rubber, cardamom, cinchona (from which quinine is extracted) and other crops are cultivated in the hills, but rain forests can still be found at higher elevations, where the annual rainfall exceeds 2,000 millimetres (79 inches).

The Western Ghats contain a huge diversity of species. About 4,000 plant species have been identified, and of these almost 75 percent are found only in the region's rain forests. These forests are also important refuges for a wide variety of animal species, including several endangered mammals – the Nilgiri langur (*Presbytis johnii*), wild dog (*Cuon alpinus*), sloth bear (*Melursus ursinus*), Malabar large-spotted civet (*Viverra megaspila civettina*), tiger (*Panthera tigris*) and gaur (*Bos gaurus*) – and at least 50 endemic species of snake.

There are eight national parks and 39 wildlife sanctuaries in the Western Ghats, covering a total area of 16,935 square kilometres (6,540 square miles) or 11 percent of the Ghats. The management status of these sanctuaries varies enormously: the Nilgiri Wildlife Sanctuary in the state of Tamil Nadu has no human inhabitants, a small number of abandoned plantations and no produce exploitation; in contrast, the Parambikulam Wildlife Sanctuary in Kerala includes large commercial teak plantations and private estates given over to the cultivation of tapioca and rice. Although some reserves in the Western Ghats are given adequate protection, many others are under threat from logging, encroachment by local people, and the creation of reservoirs. One such area is the Mundanthurai Wildlife Sanctuary in Tamil Nadu which contains extensive riverine forest where the Tambiraparani River crosses the Mundanthurai plateau. There are now plans to divert the river through a tunnel to feed the Servalar dam. Should this plan go ahead, the riverine forest would be irreparably damaged.

Sri Lanka – Most of the natural vegetation in Sri Lanka has been cleared for cultivation. This often happens in a piecemeal fashion, small plots of land being burned to make way for new vegetation.

Sundarbans Tiger Reserve – This watercourse through the Sundarbans mangrove forest reveals how the plant roots have trapped some of the silt in the Ganges, building up the level of the sediment in the delta.

Great
hornbill

Colonial life (*left, top left*) –
During the heat of the Indian
summer, the British would retire
to the hills. Hill stations
provided an escape from the
extremes of the Indian climate
from the beginning of the
summer until the end of the
monsoons in the autumn.

Nilgiri: a history of protection

The Nilgiri hills have long been known for their equable
climate. Under British occupation the region emerged as
one of India's most popular hill stations: "You may select
the temperature that you like best on these hills – Italy,
France, Devonshire or Scotland", wrote Lord Macaulay in
1834. The region also has a long history of protection, for
large areas were once kept as hunting reserves for Indian
royalty. In 1980, the decision was taken to create a
Biosphere Reserve in the Nilgiri hills and adjoining areas
of the Western Ghats.

The Nilgiri Biosphere Reserve will include at least six
existing national parks and wildlife sanctuaries, and will
cover a total of 5,510 square kilometres (2,130 square
miles). The reserve's topography is extremely varied, but
consists principally of a series of plateaux and associated
hills, ranging from 250 to 1,800 metres (820 to 5,900 feet) in
height. Vegetation types include evergreen rain forest,
montane *shola* forest with grassland, semi-evergreen
forest, moist deciduous forest and dry thorn forest.

Twenty tribal groups live in the Nilgiri Reserve. Most
notable among these are the Cholnaiks, who are the only
genuine hunter-gatherers in peninsular India. Ruins of
temples, villages and water tanks provide evidence that
the area was once inhabited by a culture which developed
an advanced system of irrigation.

The proposed Biosphere Reserve will be zoned. A core
area of 1,240 square kilometres (480 square miles) will be
used for scientific research; a buffer zone of 3,230 square
kilometres (1,250 square miles) will accommodate forestry, agriculture, animal husbandry and other uses; a
reclamation or restoration zone of 700 square kilometres
(270 square miles) will be used for research into ways of
restoring the productivity and diversity of degraded
ecosystems; and there will be a tourist zone.

More than 100 species of mammal, 550 species of bird
and 80 species of reptile and amphibian have been
recorded in the proposed Nilgiri Biosphere Reserve.
Primates include the Hanuman langur (*Presbytis entellus*)
and the bonnet macaque (*Macaca radiata*). Predators
include tiger, leopard (*Panthera pardus*), wild dog, jackal
(*Canis aureus*), striped hyena (*Hyaena hyaena*) and Indian
fox (*Vulpes bengalensis*). The Reserve's elephant (*Elephas
maximus*) population stands at around 1,500 animals.
Most important among the hoofed mammals is the Nilgiri
tahr (*Hemitragus hylocrius*) of which there are between
400 and 450 in the Reserve. Rare birds include the great
hornbill (*Buceros bicornis*) and the Ceylon frogmouth
(*Batrachostomus moniliger*). Among reptiles, the mugger
crocodile (*Crocodylus palustris*) has become rare due to
poaching and habitat loss.

Northeast India

Rain forest once covered large areas of northeast India, but as the human population has increased, thousands of square kilometres of forest have been cleared and replaced with scrub jungle. Today, about 43,000 square kilometres (16,600 square miles) of rain forest still remain. Shifting agriculture, known locally as *jhum*, is common throughout the region. It was a sustainable form of land use when the local population was much smaller, but as the population pressure has increased the situation has deteriorated. Lack of land has caused many farmers to shorten the fallow periods, and in consequence severe soil erosion can now be found over large areas. Up to 82,000 square kilometres (34,000 square miles) may be affected by *jhum*. Wood is an important source of fuel for 75 percent of India's people and its collection is another serious drain on forest resources.

The surviving forests are confined to the Assam valley, the foothills of the eastern Himalaya, and the lower parts of the Naga hills in places where the annual rainfall exceeds 2,300 millimetres (90 inches) a year. Botanically, these forests are the richest in the Indian subcontinent. They also support a great diversity of mammal and bird species.

The region covers more than 170,000 square kilometres (65,650 square miles), but includes only four national parks and three wildlife sanctuaries, which protect a total of 1,880 square kilometres (725 square miles). Plans are being developed to boost this figure to more than 9,000 square kilometres (3,475 square miles), including 17 new parks and 50 new sanctuaries.

Sri Lanka

The rain forests of Sri Lanka are situated in the southwest of the island. Only one percent of the original natural vegetation remains, mainly in the mountains. The rain forests are poorer in species than those in the Western Ghats, but are nevertheless rich, composed of about 850 endemic plant species. Some of the island's trees are so rare as to be known from only one specimen.

Because Sri Lanka is densely populated – there are 260 people per square kilometre (673 per square mile), but this is expected to rise to nearly 500 people per square kilometre (1,300 per square mile) by the year 2125 – the pressure on the rain forests is immense. Most lowland areas have already been transformed into rice fields and coconut plantations, and most hilly areas support teak plantations, tea and other crops.

Bangladesh

Once forested with mangroves, swamp forests and other tropical forests from the Ganges delta up into the hills, Bangladesh is now almost completely deforested. Less than five percent of the original rain and monsoon forest cover remains, and much of this is seriously degraded. The nation is one of the poorest countries in the world. Most people eke out a living from agriculture, and the population pressure on the land is immense, with 800 people per square kilometre (2,070 per square mile). Frighteningly, the population is expected to multiply two and a half times in the next 100 years before levelling off.

Bangladesh's largest rain forest is the Sundarbans, a vast coastal mangrove forest that covers more than 4,000 square kilometres (1,550 square miles). It grows on the silt deposited by the Ganges, Brahmaputra and Meghna Rivers. Here the tiger still survives, but the nation has already lost most of its other large mammals, including the great Indian rhinoceros (*Rhinoceros unicornis*), banteng, nilgai and swamp deer. Elephants survive only in small pockets of inland forests.

Efforts are being made to develop a national conservation strategy that can save the few remaining forests from total destruction. There has been some success in replanting mangroves in the Sundarbans to replace those already lost.

Terrorism threatens wildlife – The Manas Tiger Reserve is one of the largest and most important protected areas in Asia. It is home to 19 mammal species listed as threatened in the Indian Wildlife Protection Act. In February 1989 the reserve was invaded by Bodo extremists, part of a widespread political movement fighting for an independent state in northeastern India, and a large part of the Reserve, including the vital core area, was forcibly occupied. Thousands of trees have since been cut down for sale as timber, and the only known habitats for endangered species such as the hispid hare (*Caprolagus hispidus*) and pygmy hog (*Sus salvanius*) are now at risk. The Bodos also attacked ranger posts and offices, killing a forestry officer and at least 8 wardens. By September 1989, 30 of the 44 ranger posts and camps had been abandoned, and at least 6 of the 80 great Indian rhinoceroses in the Reserve had been killed. Hundreds of deer and wild pigs are also believed to have been shot for food.

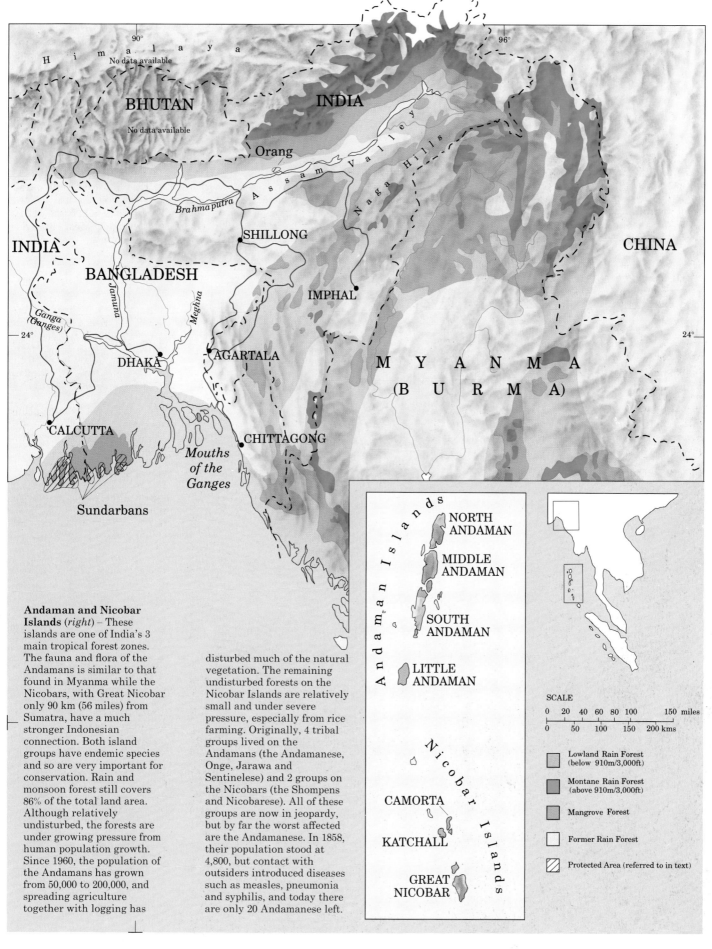

Andaman and Nicobar Islands (*right*) – These islands are one of India's 3 main tropical forest zones. The fauna and flora of the Andamans is similar to that found in Myanma while the Nicobars, with Great Nicobar only 90 km (56 miles) from Sumatra, have a much stronger Indonesian connection. Both island groups have endemic species and so are very important for conservation. Rain and monsoon forest still covers 86% of the total land area. Although relatively undisturbed, the forests are under growing pressure from human population growth. Since 1960, the population of the Andamans has grown from 50,000 to 200,000, and spreading agriculture together with logging has disturbed much of the natural vegetation. The remaining undisturbed forests on the Nicobar Islands are relatively small and under severe pressure, especially from rice farming. Originally, 4 tribal groups lived on the Andamans (the Andamanese, Onge, Jarawa and Sentinelese) and 2 groups on the Nicobars (the Shompens and Nicobarese). All of these groups are now in jeopardy, but by far the worst affected are the Andamanese. In 1858, their population stood at 4,800, but contact with outsiders introduced diseases such as measles, pneumonia and syphilis, and today there are only 20 Andamanese left.

SCALE

0   20 40 60 80 100   150 miles

0   50   100   150   200 kms

Lowland Rain Forest (below 910m/3,000ft)

Montane Rain Forest (above 910m/3,000ft)

Mangrove Forest

Former Rain Forest

Protected Area (referred to in text)

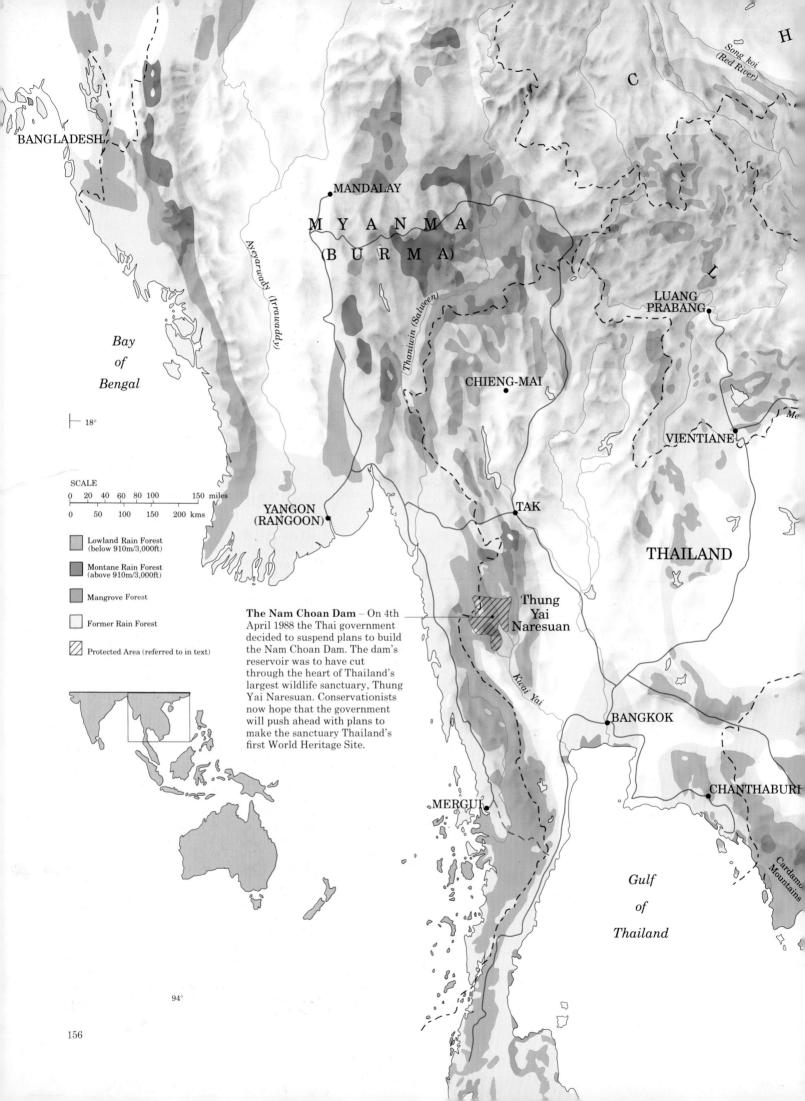

H

C

Song koi
(Red River)

BANGLADESH

● MANDALAY

M Y A N M A

(B U R M A)

L

LUANG
PRABANG ●

Bay
of
Bengal

Ayeyarwady (Irrawaddy)

Thanlwin (Salween)

CHIENG-MAI ●

├── 18°

VIENTIANE ●

Me

SCALE

| 0 | 20 | 40 | 60 | 80 | 100 | | 150 | miles |

| 0 | | 50 | | 100 | | 150 | 200 | kms |

YANGON
(RANGOON) ●

TAK ●

THAILAND

Thung
Yai
Naresuan

Lowland Rain Forest
(below 910m/3,000ft)

Montane Rain Forest
(above 910m/3,000ft)

Mangrove Forest

Former Rain Forest

Protected Area (referred to in text)

The Nam Choan Dam – On 4th
April 1988 the Thai government
decided to suspend plans to build
the Nam Choan Dam. The dam's
reservoir was to have cut
through the heart of Thailand's
largest wildlife sanctuary, Thung
Yai Naresuan. Conservationists
now hope that the government
will push ahead with plans to
make the sanctuary Thailand's
first World Heritage Site.

Kwai Yai

BANGKOK ●

CHANTHABURI ●

MERGUI ●

Cardamo
Mountains

Gulf
of
Thailand

94°

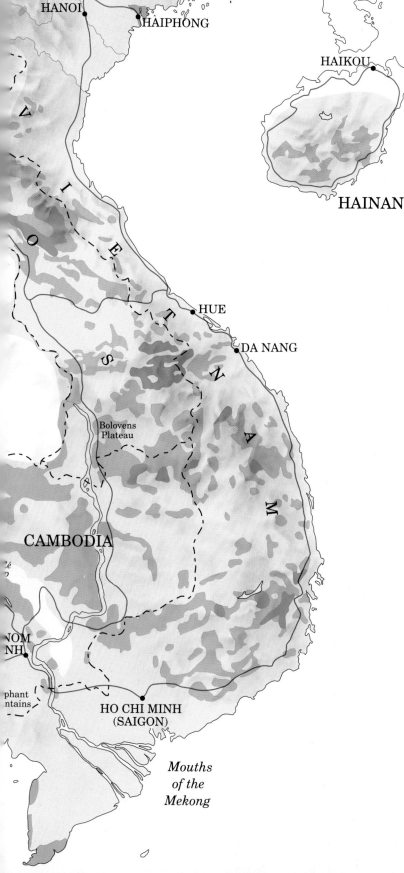

Mainland Southeast Asia

NANNING

VICTORIA
HONG KONG

HANOI

HAIPHONG

HAIKOU

HUE

DA NANG

Bolovens
Plateau

CAMBODIA

NOM
NH.

phant
ntains

HO CHI MINH
(SAIGON)

HAINAN

*Mouths
of the
Mekong*

Relic cattle – The kouprey is the
most primitive of living cattle: it
resembles species that lived
600,000 years ago. Discovered by
Western scientists only in 1937,
the kouprey lives in Thailand,
Vietnam, Laos and Cambodia. It
is now perilously close to
extinction; there are probably no
more than a few hundred animals
in the wild.

True rain forest and drier seasonal (or monsoon) forest once
covered most of Myanma (formerly Burma), Thailand, Cambodia,
Laos, Vietnam and southern China. The two types graded into
each other. The rain forest, where disturbed, developed characteristics of the monsoon forest, whereas the monsoon forest
degraded into bamboo forest, or open woodland with grasses on
the ground, particularly when subjected to regular burning.
Human disturbance of the forests of mainland Southeast Asia has
been so widespread, over such a long period of time, that it is
doubtful if any untouched rain or monsoon forest still remains.

Humans have lived in tropical Asia for up to a million years,
and great civilizations have come and gone, leaving behind traces
of their existence – for example, the twelfth-century monuments
of Angkor Wat in Cambodia are a testament to the skill and power
of the Khmer people. Originally, the inhabitants of the region
were hunter-gatherers; but with the increasing use of fire, large-
scale shifting cultivation became widespread. In the hills,
shifting cultivators cleared vast areas of forest; while along the
great rivers the once extensive freshwater swamp forests were
long ago replaced by rice paddies. The drier monsoon forests were
the first to go – the long dry season makes for better burning.
From the end of the nineteenth century, when tea, coffee, rubber
and oil palm were introduced, the rain forests came under
pressure too.

There is a long history of logging in the region. Inventories and
working plans in Myanma date back to the 1850s, when the great
teak forests began to be exploited. Right up until the end of World
War II, most of the logs were extracted with the use of elephants,
and the environmental impact was light. With the advent of
chainsaws and mechanized extraction, the damage done to the
forests has become far worse, even to the extent that the Asian
elephant (*Elephas maximus*) is now in rapid decline.

Much of the region's wildlife is now endangered as a result of
the fragmentation and degradation of the rain forests. Although
there is a very rich flora, it is also particularly vulnerable in that
perhaps one-third of the 15,000 plant species are found only in this
region. Important wild relatives of cultivated fruit trees include
ten species of mango and many citrus species. Wild relatives of
cattle are now a cause of great concern: the kouprey (*Bos sauveli*)
is close to extinction, while the gaur (*Bos gaurus*) and banteng
(*Bos javanicus*) are in decline. Birds too are suffering from the
loss of their forest habitats. Vietnam has 34 threatened species
including the Vietnam pheasant (*Lophura hatinhensis*), until
recently known only from two specimens; meanwhile Thailand
has 39 threatened species including Gurney's pitta (*Pitta gur-
neyi*), feared extinct but recently rediscovered.

Myanma

In spite of its relatively low population density of 62 people per square kilometre (24 per square mile), Myanma has one of the highest deforestation rates in the world – it is losing two percent of its rain and monsoon forests every year. The main problem in this poverty-stricken and politically isolated country is that most people depend on forest lands for their livelihoods. Shifting cultivators have been forced to clear large areas of forest through the lack of an alternative way of supporting themselves.

The rain forests are found on the west-facing slopes of the great mountain ranges, which stretch from north to south along the country's western and eastern frontiers. In general, the areas used for logging have been protected by the Forest Department and remain as forest. However, in the frontier areas there is an increasing amount of both legal and illegal logging taking place. Following the Thai government's ban on logging in 1989, many Thai entrepreneurs have turned to Myanma as a source of teak (*Tectona grandis*) and other tropical hardwoods. In February that year, 20 logging concessions were set up along the Thai border, and perhaps a further 20 by the end of the year. The initial 20 are legally allowed to export more than 200,000 tonnes (175,000 US tons) of logs to feed Thailand's timber mills, worth US$112 million a year to Myanma – a regular bonanza for a country with little export business. It seems unlikely that the Forest Department will be able to ensure that these logging concessions are exploited sustainably, and that massive degradation can in due course be avoided.

Thailand

Thailand is one of the wealthiest and most stable countries in the region. Much of its wealth was accumulated through exploitation of its own forests and other natural resources; but by 1967, Thailand was a net importer of timber. About 15 percent of the country remains covered in rain forest; a further six percent by monsoon forest – less than half of the nation's original resources. Disputes over the future of logging in the country came to a head in 1988, when 450 people were killed and hundreds more made homeless by floods caused by deforestation. A total logging ban was imposed shortly afterwards.

Thailand has a large protected area system with 59 national parks (and a further 21 proposed), 28 wildlife sanctuaries and 118 non-hunting areas. However, there are serious management problems since there is not enough money to maintain the system. Lowland forests are poorly represented and even the existing protected areas continue to be logged and used by people who rely on shifting cultivation.

Cambodia

Cambodia's fertile plains, productive fisheries and valuable forests have been exploited by several great kingdoms, including the Khmer empire. These dynasties cleared most of the forests in the centre of the country, whereas the uplands were left untouched until the 1970s, when the Vietnam war spilled over into Cambodia, leading to 20 years of disruption. Hard facts on the extent and rate of deforestation are hard to come by, but Cambodia is still suffering from the environmental effects of the Vietnam war and subsequent civil war. Between 1974 and 1978, the ruling Khmer Rouge Party systematically slaughtered most of Cambodia's educated and professional classes. Apart from the appalling loss of human life, this has resulted in a shortage of expertise in many fields, including forestry.

Extensive rain forests are still to be found in the Elephant and Cardamomes mountains, particularly on the western slopes where there are very few people. In the east, the mountains suffered extensive defoliation and bombing by the Americans during the Vietnam war because one of the main supply routes to Vietnam, the so-called Ho Chi Minh trail, passed through here. Since then, they have been further damaged by the activities of shifting cultivators.

Laos

There are rain forests in the central mountainous regions of Laos, including parts of the Bolovens plateau. In the lowlands, rain forests occur in small patches on the Mekong River plain. The greatest threat to Laotian forests is shifting cultivation; logging, mining and civil engineering projects have so far done relatively little damage. However, the improved national road network, due to be completed in the 1990s, will provide greater access for loggers and is likely to cause further deforestation. Felling has been poorly controlled in the past, and the lack of a coherent management system is likely to mean that logging will become a major threat in the future.

Laos is unusual in having no protected areas. A 47,000 square-kilometre (18,000 square-mile) system has been proposed by IUCN, including the 1,438 square-kilometre (555 square-mile) Xe Piane protected area in the south, said to contain the tiger (*Panthera tigris*), the kouprey, the Javan rhinoceros (*Rhinoceros sondaicus*) and other endangered species.

Vietnam

Almost 90 percent of Vietnam was once covered by rain and monsoon forest. By 1943, this had declined to around 43 percent. Today, less than 19 percent of the country is still forested and more than 40 percent is classified as wasteland. Extensive areas had been cleared in coastal regions and in the Mekong and Red River floodplains. From 1945 to 1975, almost uninterrupted warfare resulted in the destruction of most of the remaining forest and farmland (see box), giving rise to a new word – ecocide.

Vietnam began to establish a system of protected areas in 1962, and it is estimated that there is room for up to 87 reserves. However, the forests are so isolated that forest corridors between reserves are essential to prevent a reduction in the biodiversity.

The end of logging in Thailand?

In January 1989 a royal decree banning all logging was issued in Thailand. This pioneering move followed a disastrous landslide resulting from the expansion of government-subsidized rubber plantations, often on slopes too steep to support them. In one such case thousands of tons of mud and cut logs slid down the side of a mountain and buried several villages. Thailand was widely lauded in the international press for introducing the ban, but there is abundant evidence to suggest that illegal logging is still widespread. Small-scale cutting has increased as the large timber companies withdraw from their concessions, and thousands of people have been charged with logging offences. An additional problem is that logging companies are still being permitted to remove those logs cut, or said to have been cut, before January 1989. Many loggers anticipated the ban weeks before its introduction and in the ensuing scramble to maximize profit, carried out felling round-the-clock. Further damage to forest habitats has since resulted from the cutting of new roads in order to remove the felled trees.

The most worrying aspect of the logging ban is that it does not appear to reflect any fundamental change of attitude on the part of the Thai government towards the use of tropical hardwoods. Laos has been a major source of timber for Thailand for some years, but since the ban Thailand has hurried to sign an agreement with the Myanma government to increase the volume of wood imported. Deals were struck in order to justify the large investment in sawmills that had been made when logging was still permitted in Thailand. Therefore, in spite of the ban, the mills around Bangkok are still awash with rain forest timber.

Today nearly 80% of the Mekong delta's mangrove forest has been replanted.

Replanting Vietnam

Vietnam is facing one of its biggest challenges since the Vietnam war ended in 1975. That challenge is monumental: American and Vietnamese scientists estimate that 22,000 square kilometres (8,500 square miles) of forest and one-fifth of the country's farmland were affected as a direct result of bombing, mechanized land clearing and defoliation. The mangrove forests proved to be particularly susceptible to chemical defoliants, such as Agent Orange, and were more seriously damaged than any other forest type.

Surprisingly, more forest has been lost since 1975 than during the war. Post-war logging operations for the rebuilding of homes, schools, hospitals, roads and irrigation systems, accounted for the initial loss. Shifting cultivation, commercial logging for export, fuelwood collection and forest fires added an additional toll. Vietnam annually loses about 3,000 square kilometres (1,150 square miles) of forest.

Despite these setbacks, the Vietnamese people have begun to tackle the task of "regreening" their country with an enthusiasm that is unmatched in Southeast Asia.

Ever since Vietnam launched its National Conservation Strategy in 1986, the country has been engaged in a massive reafforestation campaign. The annual Tree Planting Tet, a lunar new year celebration started by President Ho Chi Minh, sets the entire nation in motion towards meeting its goal of planting at least 500 million trees a year. Eventually, Vietnam hopes to bring its forest cover back up to 50 percent of the country.

Vietnam's Ministry of Education has also made tree planting one of its curricular activities, requiring that every pupil plant a tree and maintain it. In 1985 and 1986, Vietnamese students planted 52 million trees and set up more than 8,100 square kilometres (3,100 square miles) of tree nurseries. Shortly after the war ended, Vietnamese scientists attempted to replant indigenous trees in areas formerly covered with moist tropical forest, but the young saplings were burned up in grass fires during the dry season. To protect the seedlings and provide them with shade, the scientists established a canopy of acacia and eucalyptus under which several species of dipterocarps are now thriving.

Peninsular Malaysia and Sumatra

The rain forests of the 11 peninsular states of Malaysia are characteristically tall in stature and rich in species. Much of the region's great biological diversity is concentrated in the lowlands: more than half of the 200 species of land mammal are found only below 330 metres (1,100 feet). As almost all Malayan wildlife depends on the forests, the lowland forests are by far the most important habitats for conservation, particularly for the larger, hoofed mammals such as the Sumatran rhinoceros (*Dicerorhinus sumatrensis*) and the Malayan tapir (*Tapirus indicus*).

Over the past 20 years there has been widespread conversion of lowland forest into settlements, plantations and other forms of cultivation. The rate of forest loss between 1986 and 1990 was estimated to be 950 square kilometres (370 square miles) per annum, or about 1.6 percent a year. Logging is relatively well managed, and logged forest provides a haven for a lot of wildlife, while some of the most important areas are protected reserves.

Sumatra is losing its rain forests faster than any other Indonesian island.

Sumatra is the second largest island in the Indonesian archipelago. It has extensive areas of lush, species-rich rain forest. Dipterocarp trees dominate the lowland forests, forming a near continuous canopy over a diversity of plants matching that found in Borneo and New Guinea. Among the most famous of the plants are the amazing *Amorphophallus titanum* (the world's tallest flower) and *Rafflesia arnoldii* (the world's broadest flower), which was originally described as "the greatest prodigy of the vegetable world". The fauna of Sumatra is one of the richest in the archipelago, and includes 196 species of mammal, 22 of which are endemic.

The wealth and productivity of the Sumatran forests – some dipterocarp forests can yield more than 100 cubic metres of excellent logs per hectare (320 cubic yards per acre) – make them particularly attractive to loggers. At 3.3 percent a year, Sumatra also has one of the highest population growth rates in Indonesia. As a result, Sumatra is losing its rain forests faster than any other Indonesian island.

Lar gibbon (*Hylobates lar*)

The importance of diet

Logged forest is often thought to be of little use in the conservation of wildlife. Although it is true that many species are adversely affected by selective logging, and some cannot survive even the slightest disturbance, a large number are able to maintain viable populations in logged forests. In the forests near Sungai Tekam in Peninsular Malaysia, the ability of a number of primates to inhabit logged areas appears to be linked to the animals' ability to change its diet. The lar gibbon (*Hylobates lar*) and the banded leaf monkey (*Presbytis melalophos*) both increase the proportion of young leaves in their diets when living in logged forest. Young leaves are particularly abundant in this habitat as a response to the increased amount of light. However, not all primate species are this flexible, and specialist feeders are particularly vulnerable. The proboscis monkey (*Nasalis larvatus*), which lives in riverine and mangrove forests in Borneo and feeds on only a few tree species, is very sensitive to any disturbance of its habitat. The orang-utan (*Pongo pygmaeus*) is also unable to tolerate the disturbance caused by logging.

Endau-Rompin: state versus government

Although conservation is a national issue in Peninsular Malaysia, relevant policies developed and implemented by the government, natural resource management and exploitation are carried out by individual state authorities. The difficulties posed by this are well illustrated by the Endau-Rompin controversy. In the early 1970s a committee, which included representatives from both the government and the state authorities, recommended the creation of a national park in the Endau-Rompin wilderness, which is shared by the states of Pahang and Johor. The importance of this wilderness lies in the fact that once existing agricultural development programmes have been fully implemented, its lowland forests will be the last refuge for some of the most threatened animal species in the region, among them the Sumatran rhinoceros and the tiger (*Panthera tigris*). In 1977, the Pahang state government granted logging rights in its section of the proposed park. This was countered by a well organized campaign, headed by the Malay Nature Society, to save the wilderness. As a result, logging ceased in 1978 and now major protected areas have been set up by both state authorities; but a national park has still not been established.

Raiding elephants – One of the major problems facing the Asian elephant (*Elephas maximus*) in Sumatra is habitat fragmentation. Logging, and the conversion of rain forest to agriculture, often leaves small numbers of elephants marooned in isolated forest islands, from which they then raid the surrounding fields. Crops such as sugar cane and rice are particularly attractive to the elephants. More recently, the Indonesian government has moved thousands of people from the islands of Java, Madura and Bali to Sumatra as part of its transmigration programme, creating villages along the edges of the forest areas where elephants are found. This has resulted in territorial conflicts between the villagers and the elephants, and today many elephants bear the scars of wounds caused by the guns used to drive them from the crops.

Isolated primates – The Mentawai group of islands has been separated from Sumatra by a deep marine trough since the middle of the Pleistocene epoch, about a million years ago. In effect, the Mentawai have been oceanic islands for about 500,000 years, their flora and fauna evolving in isolation. This has produced many endemic plants and animals, including 4 primate species: Kloss's gibbon (*Hylobates klossii*), the Mentawai leaf monkey (*Presbytis potenziani*), the Mentawai macaque (*Macaca pagensis*) and the pig-tailed langur (*Simias concolor*).

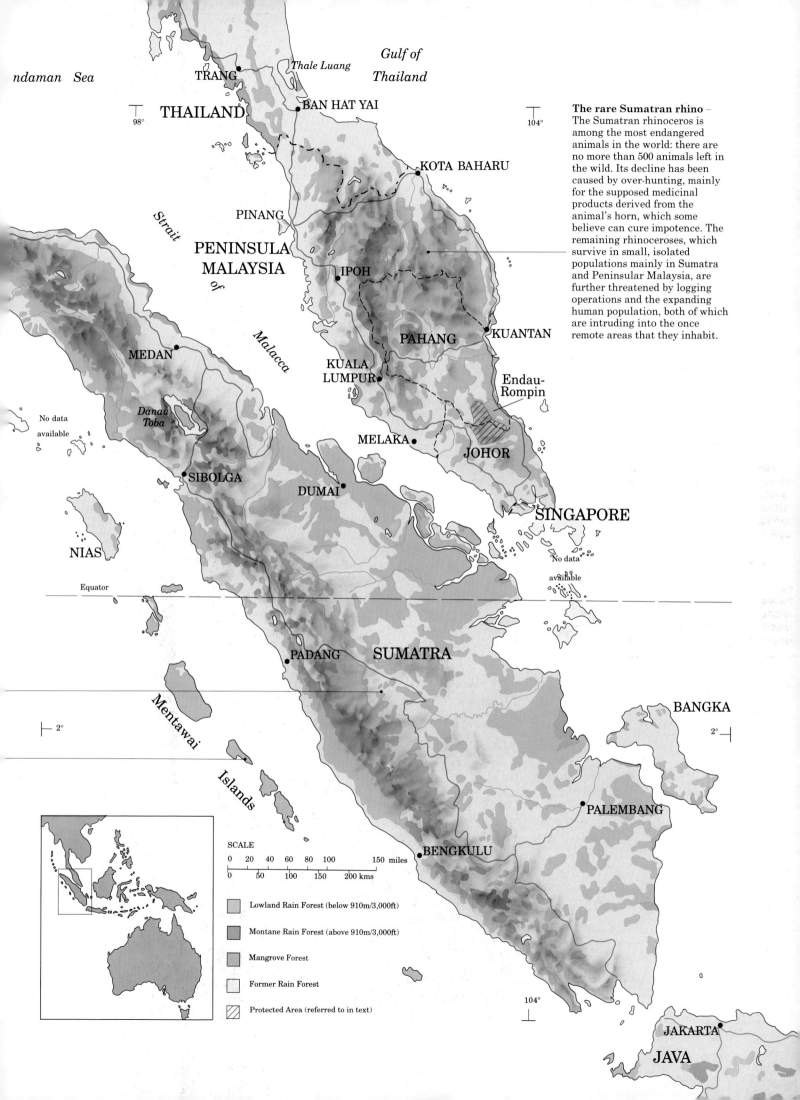

ndaman Sea

Gulf of Thailand

Thale Luang

TRANG

BAN HAT YAI

THAILAND

98°

104°

KOTA BAHARU

PINANG

PENINSULA MALAYSIA

Strait

IPOH

of

Malacca

PAHANG

KUANTAN

MEDAN

KUALA LUMPUR

Danau Toba

Endau-Rompin

SIBOLGA

MELAKA

JOHOR

DUMAI

SINGAPORE

No data available

NIAS

No data available

Equator

SUMATRA

BANGKA

Mentawai

2°

2°

Islands

PADANG

PALEMBANG

BENGKULU

The rare Sumatran rhino –
The Sumatran rhinoceros is
among the most endangered
animals in the world: there are
no more than 500 animals left in
the wild. Its decline has been
caused by over-hunting, mainly
for the supposed medicinal
products derived from the
animal's horn, which some
believe can cure impotence. The
remaining rhinoceroses, which
survive in small, isolated
populations mainly in Sumatra
and Peninsular Malaysia, are
further threatened by logging
operations and the expanding
human population, both of which
are intruding into the once
remote areas that they inhabit.

SCALE

0 20 40 60 80 100   150 miles

0   50   100   150   200 kms

Lowland Rain Forest (below 910m/3,000ft)

Montane Rain Forest (above 910m/3,000ft)

Mangrove Forest

Former Rain Forest

Protected Area (referred to in text)

104°

JAKARTA

JAVA

The Philippines and Sabah

The Philippines is an archipelago of about 7,100 islands, only 462 of which exceed 2.5 square kilometres (one square mile) in area. The two largest islands, Luzón and Mindanao, together cover 200,000 square kilometres (77,200 square miles) or 68 percent of the total land surface. There are approximately 12,000 plant species in the Philippines (3,700 of which are endemic), most of them occurring in rain forests. Many of the endemic plants have been collected by scientists only once or twice, and because recent efforts by botanists to find them in their original forest habitats have failed, some are now presumed extinct.

Whereas forests covered most of the archipelago at the beginning of this century, and still two-thirds of it in 1945, they now cover only one-fifth of the country or about 64,600 square kilometres (24,950 square miles), mostly in disturbed or degraded form. Almost all the remaining forest is in the uplands, but even this is threatened by the widespread demand that exists throughout the country for more agricultural land. The most valuable forests, both commercially and biologically, are the mature dipterocarp formations. These now cover an area of only 10,000 square kilometres (3,850 square miles) and are expected to be logged out by the year 2000.

At present, there are 59 reserves in the Philippines, covering a total area of 4,100 square kilometres (1,600 square miles). If accepted, a new plan would reduce the number of reserves to 28, but at the same time increase the total area protected to 6,450 square kilometres (2,500 square miles).

Almost all the remaining forest in the Philippines is in the uplands, but even this is threatened . . .

Montane forest on the island of Luzón

Sabah

The Malaysian state of Sabah occupies the northern tip of the island of Borneo. In 1953 more than 86 percent of the state was covered in forest, but 30 years later that proportion was down to 63 percent, and today only 41 percent or 30,000 square kilometres (11,600 square miles) is still covered. By 1980, essentially all of Sabah's productive and accessible forests had been logged, with the exception of some conservation areas. Any citizen can gain title to forested land in Sabah by clearing and working it; so as logging opened up previously inaccessible sites, settlers followed, clearing the logged forest and claiming the land. About 11,000 square kilometres (4,250 square miles) or 15 percent of Sabah is affected in this way.

Monkey-eating eagle – The Philippine eagle (*Pithecophaga jefferyi*), also known as the monkey-eating eagle, is currently found only on the islands of Luzón, Samar, Leyte and Mindanao. Despite its name, it only occasionally catches monkeys, its usual prey being a wide variety of rain forest animals and birds. The total eagle population is now estimated at fewer than 200 birds. In the past, hunting and trapping contributed to the species' decline, but deforestation is now the main threat because it is reducing suitable habitat to such an extent that breeding is being disrupted. Even the eagle's largest stronghold, in the forests of the Sierra Madre mountains in northeast Luzón, is threatened by deforestation. The Philippine Eagle Conservation Programme, started in 1969 with the support of the famous aviator Charles Lindbergh, has helped to focus international attention on the plight of this magnificent raptor.

Managing Palawan island – The island of Palawan is 425 km (265 miles) long, and has a central spine of forested mountains and narrow coastal plains, fringed by mangroves and coral reefs. The forests, although largely intact, are steadily giving way to logging, mining and uncontrolled agricultural expansion. The island is ecologically very fragile, and without careful planning there will be severe environmental degradation and impoverishment of the people living there. A recent survey concluded that a network of protected areas would not be sufficient to prevent environmental deterioration, mainly because it would not receive the support of local communities. Instead, it was proposed that a graded system of protective management should be introduced over the whole of Palawan. Some areas would be under strict control (to protect watersheds, preserve biodiversity and protect tribal peoples), while others would be left relatively unregulated. There would also be areas dedicated to research, tourism and recreation. In this way the local community would be involved at all levels of the management system.

MALAYSIA

BRUNEI

SARAWAK

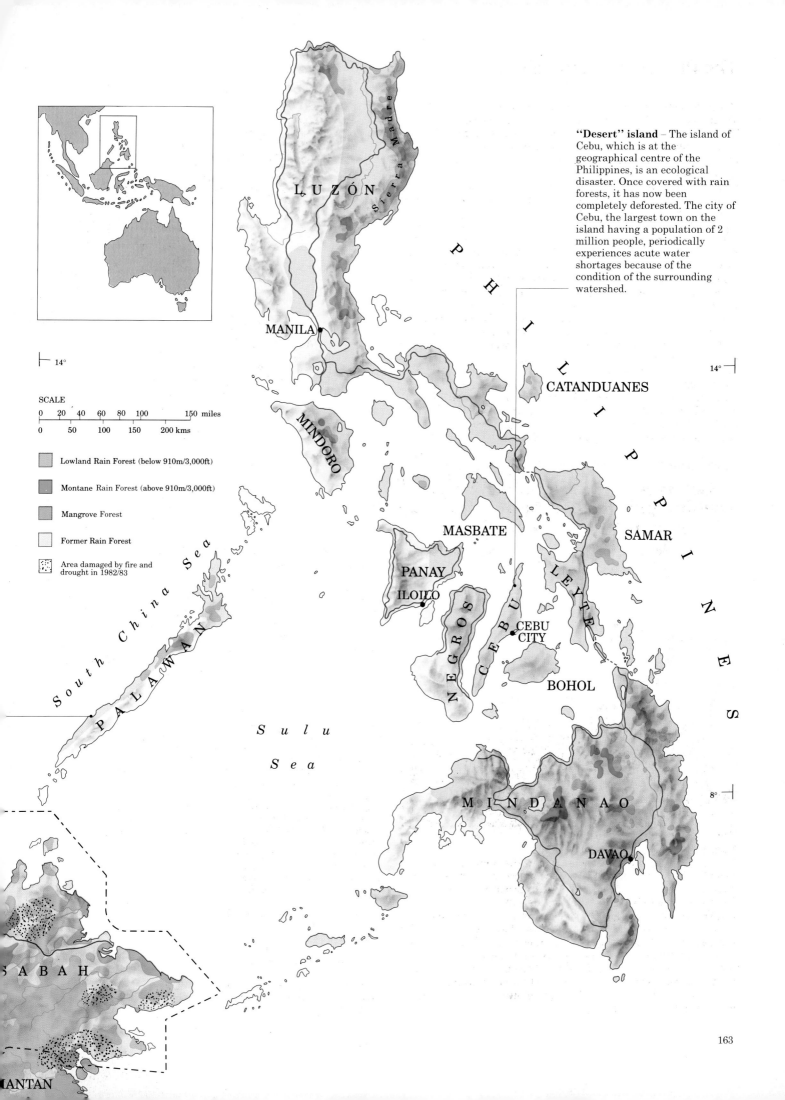

"Desert" island – The island of Cebu, which is at the geographical centre of the Philippines, is an ecological disaster. Once covered with rain forests, it has now been completely deforested. The city of Cebu, the largest town on the island having a population of 2 million people, periodically experiences acute water shortages because of the condition of the surrounding watershed.

14°

SCALE

| 0 | 20 | 40 | 60 | 80 | 100 | | 150 | miles |
| 0 | | 50 | | 100 | 150 | | 200 | kms |

Lowland Rain Forest (below 910m/3,000ft)

Montane Rain Forest (above 910m/3,000ft)

Mangrove Forest

Former Rain Forest

Area damaged by fire and drought in 1982/83

LUZÓN

Sierra Madre

MANILA

MINDORO

CATANDUANES

PHILIPPINES

MASBATE

SAMAR

PANAY

ILOILO

NEGROS

CEBU

LEYTE

CEBU CITY

BOHOL

South China Sea

PALAWAN

Sulu Sea

MINDANAO

DAVAO

SABAH

ANTAN

14°

8°

163

Mt Apo – One of the few remaining montane rain forests in the Philippines is found on the slopes of Mt Apo on the island of Mindanao. The inaccessibility of the region has been one of the major factors in its survival.

Red-vented cockatoo

The multipurpose rattan – Rattans are climbing palms which serve as the main raw material in the manufacture of cane furniture. They are also used to make fish traps, sleeping mats, hammocks, hats, walking sticks, toothbrushes and twine. The young shoots of most Philippine species are edible. The apexes of 2 species, *Daemonorops melanochaetes* and *Daemonorops hallieriana*, are cooked with fish and coconut milk and served as a dish for important guests. The juice from the fruits of the rattan may be used as a dye or as medicine for the treatment of rheumatism, asthma, snake bites and various intestinal disorders. There are about 69 species of rattan in the Philippines, and of these 10 are commercially harvested. The local rattan industry employs about 10,000 people. But as a result of over-harvesting, it has now become necessary for the Philippines to import raw, unworked rattan poles from other Southeast Asian countries. In the past, rattan products were an important source of foreign exchange, but the recent ban on cane exports by Indonesia has caused serious shortages of raw materials.

Birds of the Philippines

Of the 540 or so bird species found in the Philippines, 388 breed there, 119 are migrants passing through, and 34 are irregular visitors. Most of the breeding species originally came from Malaysia; a few arrived instead from China, Sulawesi, Maluku and New Guinea. At least 162 of these are endemic. Most of the birds live in the forests and are sensitive to any disturbance, whether from logging or slash-and-burn cultivators. For example, a recent survey on the island of Cebu found that only one of the original ten forest species had survived the logging in the area. In addition, many birds, including pigeons, doves and horn-bills, are caught for food. Yet others are trapped for the caged bird market. The red-vented cockatoo (*Cacatua haematuropygia*) and various members of the parrot family have been particularly badly affected.

It is estimated by the International Council for Bird Preservation that as many as 34 species may be immediately threatened by habitat destruction, hunting or trapping. Some of these species, such as the Negros fruit dove (*Ptilinopus arcanus*) and Mindoro scops owl (*Otus mindorensis*), have only ever been found in the localities where they were originally discovered. Others, including the Mindoro imperial pigeon (*Ducula mindorensis*), have not been seen by ornithologists for many years.

Japan's search for hardwoods

Japan first imported Philippine "mahogany" (actually timber from dipterocarp trees) in 1951. These imports were intended to complement domestic hardwoods for plywood manufacture. Logging to meet Japan's need increased steadily throughout the 1950s and, driven by profits and mechanized harvesting, the Philippine timber boom continued during the 1960s, peaking in 1969. As the boom gained momentum, the government was unable to supervise concessions effectively or to enforce logging regulations. Links between timber companies and politicians further eroded government control. In the early 1970s exports started to decline after the more accessible dipterocarp forests had been exploited; the remaining sources were not only more expensive to log but also of a lower quality. At the same time, heightened conservation awareness in the Philippines prompted an initiative to curb timber exports through a variety of forest protection ordinances. In 1976 logging was banned in parts of Luzón, Catanduanes, Masbate, Leyte and Negros as well as on all small islands; but effective implementation of the law has been limited by short-term political considerations, and illegal timber smuggling continues to be a problem.

As the supply of hardwoods from the Philippines declined, the shortfall was taken up by Indonesia, as the province of Kalimantan was opened up to logging. By 1971, Indonesia had already replaced the Philippines as Japan's largest log supplier. Within three years Indonesia was supplying 47 percent of Japanese tropical log imports. Many Japanese trading companies had entered into joint ventures with Indonesian concerns, just as they had done

a decade earlier in the Philippines. But the logging in Indonesia was on a grander scale because it had been planned in advance, as a result of cooperation between the Japanese government and timber corporations.

Malaysia had taken over from Indonesia as Japan's main supplier of tropical logs as early as 1978. During that year, Malaysia exported 10.7 million cubic metres (380 million cubic feet) of hardwood to Japan, which represented 49 percent of Japanese tropical log imports.

Malaysia can be divided into three main timber producing regions. In Peninsular Malaysia there is a near total ban on log exports, so timber exports consist mostly of sawn wood. All Malaysian log supplies come from the Malaysian states of Sabah and Sarawak, which have their own forestry departments and logging regulations. Japanese imports from Sabah peaked in 1978 at 9.2 million cubic metres (325 million cubic feet). But this level was not sustained for long and by 1987 imports had fallen to around seven million cubic metres (250 million cubic feet).

Today, Japan still gets half of its hardwood logs from Sabah. Much of the rest comes from Sarawak, whose exports to Japan have been growing steadily and have now reached 5.5 million cubic metres (195 million cubic feet) a year. The authorities in Sabah and Sarawak have repeatedly pledged that they will continue to supply Japan with a steady flow of logs. Japanese traders are nevertheless constantly on the lookout for new sources of supply. Other Asian countries currently supplying Japan with logs include Papua New Guinea, the Solomon Islands, Myanma (Burma) and Vietnam.

(*Above*) Much of the imported tropical timber ends up on waste tips.

(*Above*) Tropical woods are often used in the construction of new homes.

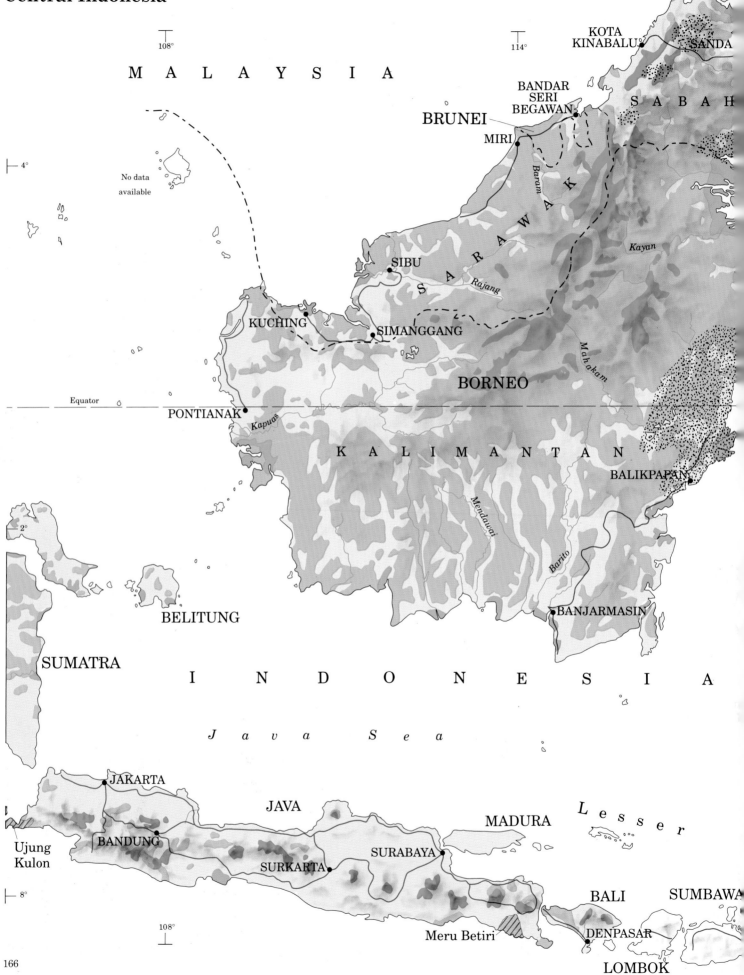

Central Indonesia

MALAYSIA

KOTA
KINABALU
SANDA

BANDAR
SERI
BEGAWAN

BRUNEI

S A B A H

MIRI

4°

No data
available

Baram

S
A
R
A
W
A
K

Kayan

SIBU

Rajang

KUCHING

SIMANGGANG

BORNEO

Mahakam

Equator

PONTIANAK

Kapuas

K A L I M A N T A N

2°

BALIKPAPAN

Mendawai

BELITUNG

Barito

SUMATRA

BANJARMASIN

I N D O N E S I A

J a v a   S e a

L e s s e r

JAKARTA

JAVA

MADURA

BANDUNG

Ujung
Kulon

SURABAYA

SURKARTA

8°

BALI

SUMBAWA

Meru Betiri

DENPASAR

108°

LOMBOK

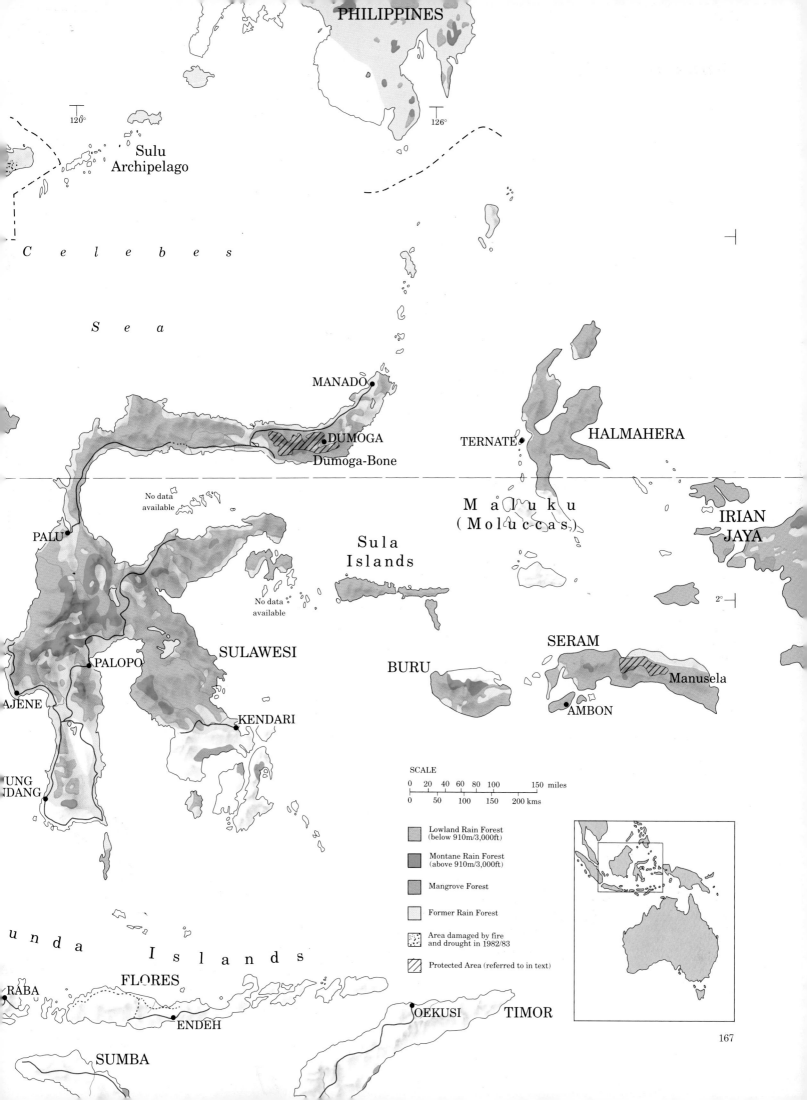

PHILIPPINES

120°   126°

C e l e b e s

Sulu
Archipelago

S   e   a

MANADO

DUMOGA
Dumoga-Bone

TERNATE   HALMAHERA

M a l u k u
(M o l u c c a s)

IRIAN
JAYA

No data
available

S u l a
I s l a n d s

PALU

2°

No data
available

SERAM

Manusela

BURU

SULAWESI

PALOPO

AJENE

KENDARI

AMBON

UNG
DANG

SCALE

0   20 40 60 80 100   150 miles

0   50   100   150   200 kms

Lowland Rain Forest
(below 910m/3,000ft)

Montane Rain Forest
(above 910m/3,000ft)

Mangrove Forest

Former Rain Forest

Area damaged by fire
and drought in 1982/83

Protected Area (referred to in text)

u n d a   I s l a n d s

FLORES

RABA

OEKUSI   TIMOR

ENDEH

SUMBA

167

The Indonesian archipelago with its 13,667 islands is one of the most biologically valuable places on earth. Nearly ten percent of the world's rain forest and almost 40 percent of all the rain forest in Asia is to be found there. Indonesia covers only 1.3 percent of the world's land surface, but it is one of the richest areas of endemism and species diversity, having more than 500 mammal species (nearly 200 of which are endemic), 1,500 bird species (17 percent of the world's total avifauna), 7,000 species of fish, 1,000 species of reptile and amphibian, tens of thousands of invertebrates and more then 20,000 plant species, including more than 10,000 tree species.

Despite its global significance, Indonesia is second only to Brazil in the rate at which its forests are being converted to agricultural and other uses. Recent estimates have put the rate of deforestation at 7,000 square kilometres (2,700 square miles) a year, but the real figure may be as high as 12,000 square kilometres (4,600 square miles). Nonetheless, extensive rain forests still survive on all the large islands except Java, and Indonesia as a whole still has more than 1.1 million square kilometres (425,000 square miles) of rain forest. Increasingly, these forests are being disturbed by logging (which can be very destructive in this part of the world), shifting cultivation (which is believed to affect 112,000 square kilometres (43,000 square miles) in Kalimantan alone), and the government-sponsored development of tree plantations and transmigration programme.

The Indonesian transmigration programme (see page 41) is the world's largest programme for voluntary, government-assisted migration. However, dramatic reductions in the Indonesian budget, resulting from the fall in oil prices in the mid-1980s, have led to a virtual standstill in the movement of new transmigrants since 1987. This lull has provided an unexpected opportunity to improve the planning of future transmigration.

Central Indonesia can be divided into five biogeographical areas: Java, Kalimantan, the Lesser Sunda Islands, Sulawesi and the Moluccas (also known as Maluku).

Java
Java is one of the most densely populated islands in the world. More than 90 percent of its natural vegetation has been destroyed, and most of the remaining primary forest is now found only in remote mountainous regions above 1,400 metres (4,600 feet). Virtually all the lowland rain forest has been cleared for farming and tree plantations, but a few isolated fragments remain, the largest being along the southeast coast.

Despite such massive deforestation, the island of Java still boasts 50 surviving Javan rhinoceroses (*Rhinoceros sondaicus*) in the Ujung Kulon National Park. However, tigers (*Panthera tigris*) may now be extinct in their last stronghold, the Meru Betiri Reserve in the east of the island.

Lesser Sunda Islands
The rain forests of the Lesser Sunda Islands (also known as Nusa Tenggara) are far less lush than those in the rest of Indonesia. The

region has a low annual rainfall, which makes the rain forests particularly vulnerable to fire during the dry season.

Although the population pressure is not as high as in other parts of the Indonesian archipelago – the Lesser Sunda Islands has 70 people per square kilometre (27 per square mile) – the forest is often burned (often for no obvious reason) and this has caused massive deforestation.

Kalimantan
Kalimantan, which covers the southern half of the island of Borneo, supports one of the largest expanses of tropical rain forest in Southeast Asia. The region contributed more than any other part of Indonesia to the US$2.5 billion that the country gained from exporting timber in 1987, and it continues to do the same today. Much of the forest has been heavily disturbed or degraded, and the roads made by the loggers tend to be used by shifting cultivators.

Between September 1982 and July 1983 more than 40,000 square kilometres (15,400 square miles) of tropical forest on the island of Borneo was devastated by drought and fire. Of this area, some 83 percent belonged to Kalimantan. The conflagration is believed to have been started by shifting cultivators who were unaware of the risk of using fire to clear land during a drought. It was able to spread quickly into selectively logged areas, where dry combustible material from tree extraction littered the forest floor, and also into peat swamps, where the dried surface soil burned fiercely. Fire is a natural part of the ecosystem and there are already signs of recovery, but full regeneration will take a very long time because of the scale of the fire (which was not immediately appreciated by local authorities).

Kalimantan's fauna contains at least 40 endemic mammals, mostly among the bats and rodents, and is famous for its populations of orang-utans (*Pongo pygmaeus*), proboscis monkeys (*Nasalis larvatus*), gibbons (*Hylobates* spp.) and other primates, its avifauna which includes such colourful birds as hornbills, pittas and barbets, as well as 30 endemic species, and its reptile and fish fauna, thought to be the richest in the region.

Sulawesi
On the island of Sulawesi there are still large areas of primary rain forest, despite the fact that shifting cultivators have cleared significant areas in the south and some parts of the north of the island. In 1980 about 55 percent of Sulawesi was forested and today the forest cover per inhabitant is still greater than in Sumatra, Java or the Lesser Sunda Islands. However, of all Indonesia's principal islands, Sulawesi is the one most threatened by the effects of rapid development. Between 1977 and 1980, almost 40,000 migrants arrived on Sulawesi from Java as part of Indonesia's transmigration programme, and these migrants now occupy more than 4,700 square kilometres (1,800 square miles) of former rain forest. The Dumoga-Bone National Park is the most important conservation area in Sulawesi. More than 90 percent of it is covered in rain forest at altitudes between 500 and 2,000 metres (1,600 and 6,500 feet), and all of Sulawesi's protected mammals are present. The park is also extremely rich in birds and, although detailed surveys have yet to be carried out, 170 species have already been recorded.

The Moluccas
The Moluccas is an archipelago of hundreds of islands ranging greatly in size. The largest tracts of rain forest are found on the islands of Halmahera and Seram. But timber concessions have been granted on both these islands, and much of the lowland forest has already been disturbed.

The fauna of the Moluccas shows a very high level of endemism among birds and mammals, particularly among the bats and rodents, and contains many larger mammals including deer, monkeys, civets and phalangers. The avifauna includes about 460 species of birds, some 30 percent of which are also endemic.

After a pig hunt, oven stones are heated to cook the meat.

Chick emerging after underground hatching.

Rotting incubators

Megapodes are mound-building birds which do not depend on their own body heat to incubate their eggs or young. Instead, they use the heat generated by rotting vegetation to do the job. Standing on one leg and scratching backwards with the other, megapodes rake together a huge heap of vegetation on the forest floor, piling up a roughly conical mound about 1.5 metres (five feet) high and about six metres (20 feet) in diameter. After the vegetation has started to rot and generate heat, the nest is ready. The female then climbs on the mound, digs a hole, and deposits a single egg. She covers the hole up again, leaves the nest and returns the next day to lay another egg. Each female lays from five to eight pale-pink eggs.

Pork hordes after moveable feast

A key feature of the rain forest of inland Borneo is their "fruiting seasons", during which many different trees produce their fruit and seed at the same time. These seasons are probably triggered by an external cue, such as a water shortage, and they tend to be quite short and unpredictable, with no guarantee that one will happen in any particular year or at any particular time of the year. During one such season, the trees in different localities tend to fruit at slightly different times, creating a moveable feast. A consequence of this is that the fruit and seed eaters of the forest have difficulty in matching their population levels to the food supply. Under these circumstances they tend to become nomadic, or to be highly mobile within an large home range.

The bearded pig (*Sus barbatus*), which is the only wild pig in Borneo, has a very diverse diet including roots, fungi, soil insects and rotting wood, small vertebrates and carrion. It also eats fallen fruit produced by a number of tree species, especially oaks, chestnuts and dipterocarps. During a fruiting season large herds of pigs move through the forest in search of ripe fruit. Such activity was well documented in the upper Baram River area of Sarawak in 1983, 1984 and 1985, and similar patterns were subsequently reported by local residents during 1986 and 1987. In 1983 the travelling pig population probably exceeded one million animals. This horde of pork was extensively plundered. The people of the Baram are heavily armed and enthusiastic hunters. In a sample of 581 families, each possessed on average two spears or spear-blowpipes and three hunting dogs, and every other family had a shotgun. During 1983, as the pigs crossed the Baram and Silat Rivers, each of the 577 families on these rivers killed an average of 33 pigs. Allowing for the killing of stragglers outside the main migration, at least 20,000 pigs were slaughtered as the migrating wave-front rolled over the Baram and Silat in 1983, or roughly eight percent of the travelling population.

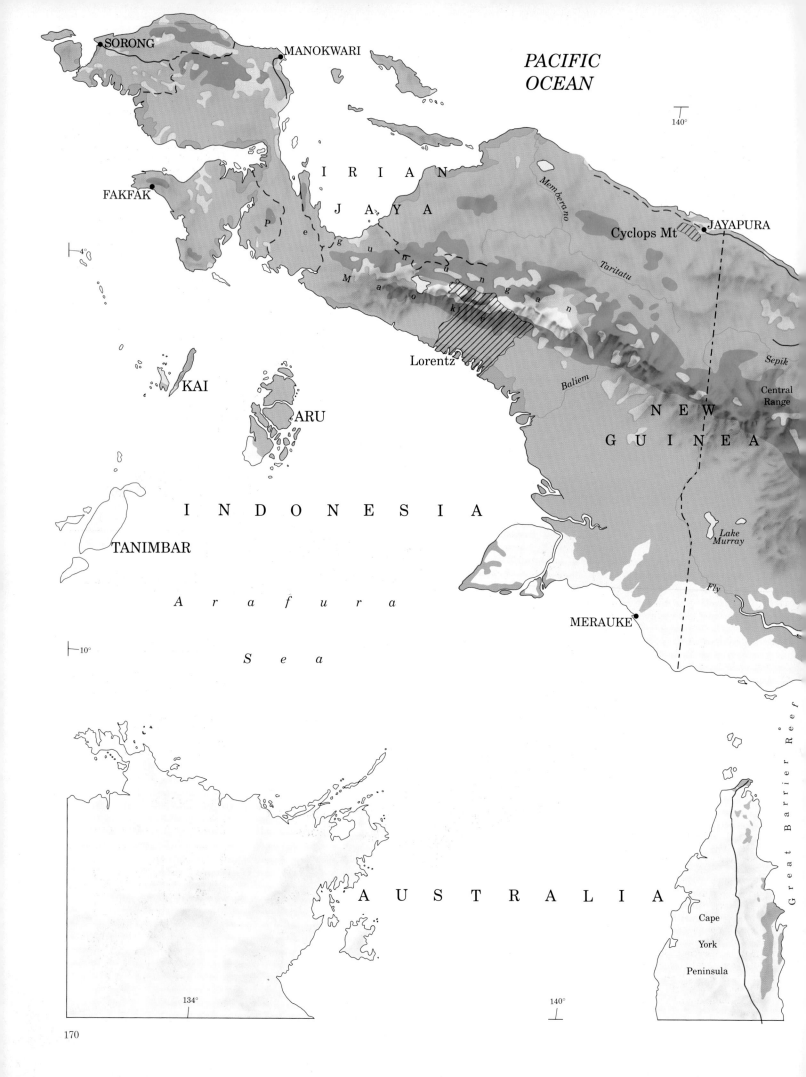

PACIFIC
OCEAN

SORONG

MANOKWARI

140°

I R I A N

J A Y A

Memberano

Cyclops Mt

JAYAPURA

FAKFAK

P

e

Taritatu

n

g

u

u

r

n

4°

M

a

o

n

g

a

n

k

e

Lorentz

Sepik

Baliem

Central
Range

KAI

N E W

ARU

G U I N E A

I N D O N E S I A

TANIMBAR

Lake
Murray

A r a f u r a

Fly

10°

S e a

MERAUKE

Great Barrier Reef

A U S T R A L I A

Cape

York

Peninsula

134°

140°

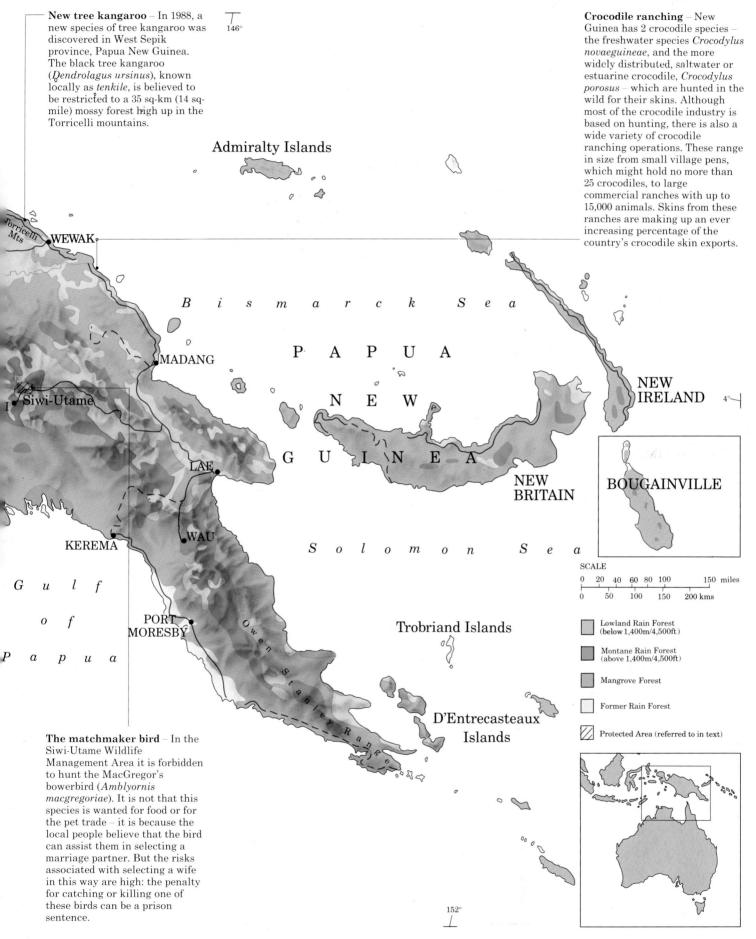

New Guinea

New tree kangaroo – In 1988, a new species of tree kangaroo was discovered in West Sepik province, Papua New Guinea. The black tree kangaroo (*Dendrolagus ursinus*), known locally as *tenkile*, is believed to be restricted to a 35 sq-km (14 sq-mile) mossy forest high up in the Torricelli mountains.

Crocodile ranching – New Guinea has 2 crocodile species – the freshwater species *Crocodylus novaeguineae*, and the more widely distributed, saltwater or estuarine crocodile, *Crocodylus porosus* – which are hunted in the wild for their skins. Although most of the crocodile industry is based on hunting, there is also a wide variety of crocodile ranching operations. These range in size from small village pens, which might hold no more than 25 crocodiles, to large commercial ranches with up to 15,000 animals. Skins from these ranches are making up an ever increasing percentage of the country's crocodile skin exports.

The matchmaker bird – In the Siwi-Utame Wildlife Management Area it is forbidden to hunt the MacGregor's bowerbird (*Amblyornis macgregoriae*). It is not that this species is wanted for food or for the pet trade – it is because the local people believe that the bird can assist them in selecting a marriage partner. But the risks associated with selecting a wife in this way are high: the penalty for catching or killing one of these birds can be a prison sentence.

146°

Admiralty Islands

Torricelli Mts

WEWAK

B i s m a r c k   S e a

MADANG

Siwi-Utame

I

LAE

WAU

KEREMA

G u l f

o f

P a p u a

PORT MORESBY

Owen Stanley Range

P A P U A

N E W

G U I N E A

NEW BRITAIN

NEW IRELAND

4°

BOUGAINVILLE

S o l o m o n   S e a

Trobriand Islands

D'Entrecasteaux Islands

SCALE

| 0 | 20 | 40 | 60 | 80 | 100 | | 150 | miles |

| 0 | | 50 | 100 | 150 | | 200 | kms |

Lowland Rain Forest (below 1,400m/4,500ft)

Montane Rain Forest (above 1,400m/4,500ft)

Mangrove Forest

Former Rain Forest

Protected Area (referred to in text)

152°

The island of New Guinea contains the largest expanse of rain forest in Southeast Asia – some 700,000 square kilometres (270,000 square miles) in total – and the world's largest reserves of the sago palm (*Metroxylon sagu*). About 80–85 percent of the forest in Irian Jaya, the western half of the island of New Guinea, and 75–80 percent of that in Papua New Guinea is still undisturbed, and vast areas of the interior have yet to be explored. In terms of plant diversity and species endemism in Southeast Asia, New Guinea is second only to Borneo.

The major vegetation types are mangrove, swamp forest, lowland and montane rain forest, eucalyptus woodland, alpine shrub and savanna grassland. Unlike the forests of Kalimantan and Sumatra, dipterocarps are not the dominant trees in the rain forest canopy; instead, other large timber trees are common, including *Intsia* spp. and *Calophyllum* spp. in the lowlands, and *Agathis* spp. and *Araucaria* spp. at higher altitudes. The mammalian fauna contains some Australian elements, including several marsupials such as phalangers and tree kangaroos. New Guinea is also famous for its splendid avifauna, in particular its birds of paradise, crowned pigeons, cassowaries, cockatoos and parrots.

Because 97 percent of the land is controlled by local communities, the development of conventional national parks and reserves has proved difficult. Instead, wildlife management areas have been introduced in an attempt to promote conservation by using traditional methods of resource management. In these areas the land remains in the possession of the customary landowners who form their own management committees to protect the wildlife within their territory. In 1986, there were 12 wildlife management areas, occupying 6,800 square kilometres (2,600 square miles) of land and approximately 1,100 square kilometres (425 square miles) of marine and coastal resources.

There have always been traditional methods of environmental management and conservation. A mixture of traditional rules and taboos have helped to protect certain species and places. Temporary bans on hunting and fishing often prevented over-exploitation and ensured sustainable yields. Access to particular sites, such as the trees in which birds of paradise display, was guarded from outsiders. These forms of traditional conservation continue to exist in many parts of Papua New Guinea, but as human pressure increases, new protective measures are needed.

In terms of plant diversity and species endemism in Southeast Asia, New Guinea is second only to Borneo.

Irian Jaya

The easternmost province of Indonesia, Irian Jaya, shares a common border of 736 kilometres (457 miles) with Papua New Guinea. It has roughly 350,000 square kilometres (135,000 square miles) of rain and monsoon forest. The rain forest covers Irian Jaya with the exception of the southeast of the province, which lies in the rain shadow of the Pegunungan Maoke mountains and supports drier savanna woodlands. Although the human population of Irian Jaya is still relatively small (1.1 million people), there is increasing pressure on the land to use it for agriculture (as part of Indonesia's transmigration programme) and timber and mineral extraction. The Indonesian government is well aware of the need to establish a network of reserves before a situation arises in which such areas have to be reclaimed from damaged forest. The first design for such a network was made in 1978, but little of this has been implemented. During 1984–85, the World Wide Fund for Nature and the Indonesian government produced the first management plan for a protected area in Irian Jaya – the 225 square-kilometre (87 square-mile) Cyclops Mountain Nature Reserve, which overlooks the provincial capital Jayapura, protecting its watershed. Developed in collaboration with local people, the Cyclops plan called for the development of conservation zones, ranging from a highly protected core zone to various degrees of traditional land uses in a number of buffer zones. Work to implement the plan is still continuing. Another very important reserve, and much bigger than Cyclops, is the Lorentz Nature Reserve, which extends over 21,500 square kilometres (8,300 square miles). No other reserve in Indonesia covers such a range of altitudes and habitats.

Papua New Guinea

Like Irian Jaya, Papua New Guinea still has more than three-quarters of its original forest cover, about 350,000 square kilometres (135,000 square miles). Deforestation is slow, but shifting cultivation in the highlands and logging, plantation development and mining in the lowlands are increasing the rate of loss. Shifting cultivation now covers about 10,000 square kilometres (3,900 square miles).

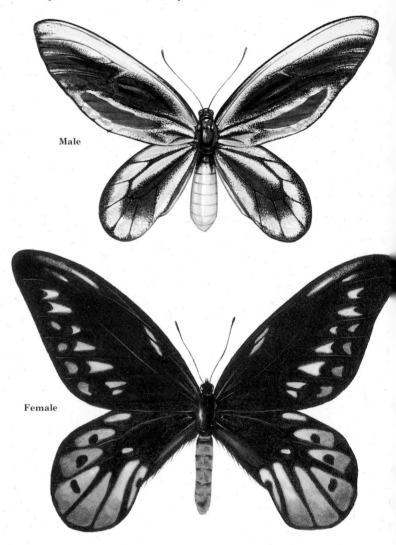

Male

Female

Big enough to shoot – Queen Alexandra's birdwing (*Ornithoptera alexandrae*) is the world's largest butterfly. The females of the species can have wingspans in excess of 25 cm (10 in). In fact, because of its size and very high flight, the first specimen was brought down by a shotgun blast. Queen Alexandra's birdwing butterfly is found only in a very small region in southeastern Papua New Guinea, and even there it is rare. Its habitats are threatened by the expansion of oil-palm plantations.

. . . although only four million people live on New Guinea, there are at least 1,000 different languages and dialects . . .

The many peoples of New Guinea

A recent estimate suggests that although only four million people live on New Guinea, there are at least 1,000 different languages and dialects spoken on the island. With very few exceptions, the inhabitants of New Guinea are horticulturalists, cultivating coconuts, yams, taro, bananas and a number of other food plants. Fish are caught with nets, spears, and occasionally hooks, and the dugong (*Dugong dugong*) is hunted by some coastal tribes using harpoons. On land, pigs, both domesticated and wild, are the main source of meat, although wallabies and many birds are also eaten. Betel-nut is chewed with lime and pepper plant, and the use of tobacco is widespread.

Apart from on some islands such as the Trobriands, New Guinean societies are all without any form of chieftainship or inherited rank, and a very aggressive individualism predominates. There is constant competition for prestige among adult men, and each is judged according to his achievements. The political units are small, usually a single village or, in parts of the highlands, a dispersed set of hamlets.

Fighting and warfare are common throughout the island. The Jale people who inhabit the highlands of New Guinea, east of the Baliem valley, accept war as part of their way of life. Wars are waged mercilessly against neighbours in the same valley and even against settlements in other regions. Hostilities may last as long as a generation – although interspersed with periods of truce – or peter out in a single day. Although deaths are limited by

the taboos that surround all Jale conflicts, the dead may still be subjected to further spite. The supreme act of Jale vengeance is to eat the bodies of slain foe.

Some New Guinean tribes are among the most isolated in the world. On the southwest coast of the island, a vast rain forest, dissected by a thousand rivers, reaches to the edge of the Arafura Sea. This area is occupied in part by a group of people known as the Asmat. The Asmat are rarely visited by outsiders, and their culture sharply reflects both this isolation and the uncompromising nature of their environment. Much of their territory is covered by swamp forest, which is subject to periodic flooding. Stones are a rarity in this environment, and so the few stone axes that the Asmat possess are obtained from the highland tribes by barter. Pottery is unknown, and all food is roasted over an open fire.

The basis of the Asmat's culture are the trees that surround them. The sago palm is not only the main source of food, but also the most basic raw material. Houses, hunting implements, canoes and paddles are all made from the wood of this tree. Furthermore, to the Asmat, tree and human are symbolically identical: the fruit of a tree is used as a metaphor for a man's head, which has profound implications for the Asmat – they are headhunters.

Fruit-eating animals symbolize this concept, and birds, including the palm cockatoo (*Proboseiger aterrimus*) and hornbills, are both regularly honoured in Asmat paintings and decorations.

Australia

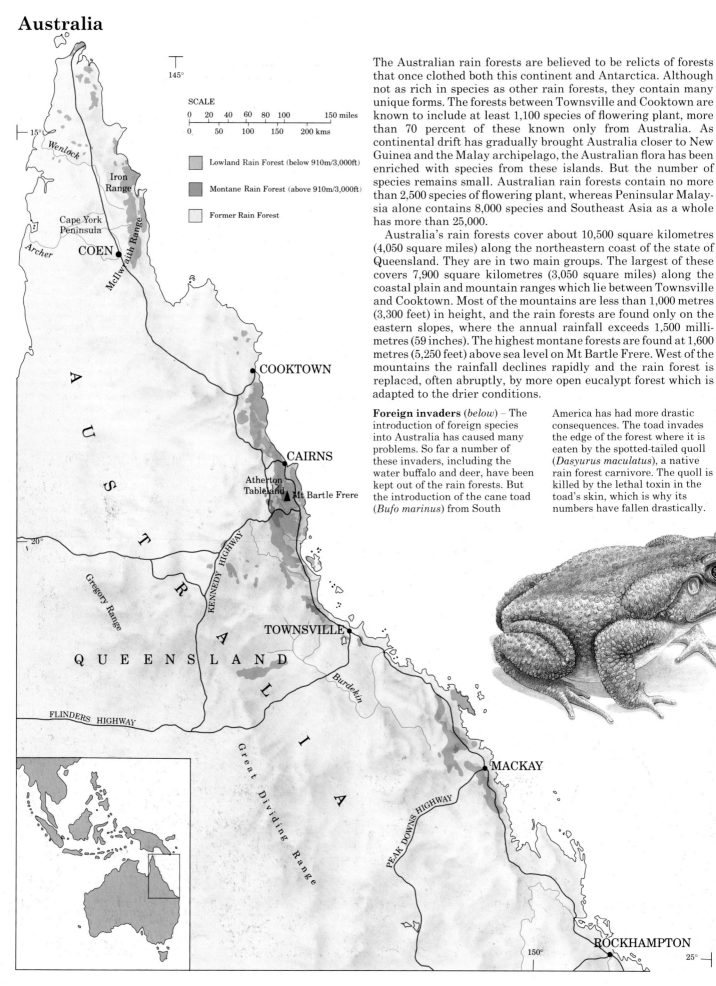

SCALE

0	20	40	60	80	100		150 miles

0	50	100	150	200 kms

Lowland Rain Forest (below 910m/3,000ft)

Montane Rain Forest (above 910m/3,000ft)

Former Rain Forest

The Australian rain forests are believed to be relicts of forests that once clothed both this continent and Antarctica. Although not as rich in species as other rain forests, they contain many unique forms. The forests between Townsville and Cooktown are known to include at least 1,100 species of flowering plant, more than 70 percent of these known only from Australia. As continental drift has gradually brought Australia closer to New Guinea and the Malay archipelago, the Australian flora has been enriched with species from these islands. But the number of species remains small. Australian rain forests contain no more than 2,500 species of flowering plant, whereas Peninsular Malaysia alone contains 8,000 species and Southeast Asia as a whole has more than 25,000.

Australia's rain forests cover about 10,500 square kilometres (4,050 square miles) along the northeastern coast of the state of Queensland. They are in two main groups. The largest of these covers 7,900 square kilometres (3,050 square miles) along the coastal plain and mountain ranges which lie between Townsville and Cooktown. Most of the mountains are less than 1,000 metres (3,300 feet) in height, and the rain forests are found only on the eastern slopes, where the annual rainfall exceeds 1,500 millimetres (59 inches). The highest montane forests are found at 1,600 metres (5,250 feet) above sea level on Mt Bartle Frere. West of the mountains the rainfall declines rapidly and the rain forest is replaced, often abruptly, by more open eucalypt forest which is adapted to the drier conditions.

Foreign invaders (*below*) – The introduction of foreign species into Australia has caused many problems. So far a number of these invaders, including the water buffalo and deer, have been kept out of the rain forests. But the introduction of the cane toad (*Bufo marinus*) from South America has had more drastic consequences. The toad invades the edge of the forest where it is eaten by the spotted-tailed quoll (*Dasyurus maculatus*), a native rain forest carnivore. The quoll is killed by the lethal toxin in the toad's skin, which is why its numbers have fallen drastically.

The second major group of rain forests, 2,600 square kilometres (1,000 square miles) in extent, lies further north and is separated from the first group by an area of eucalypt forest. It consists of scattered patches of forest on the Cape York Peninsula. The largest of these lies between the McIlwraith Ranges and the Iron Range, just north of Coen. Smaller patches are found in the headwaters of the Jardine River and on the northern tip of the Cape York Peninsula.

Assessments of how much of the original rain forest has been cleared vary widely. Australian scientists estimate that 81 percent of the southern group (Townsville to Cooktown) and 99.5 percent of the northern group (Cape York Peninsula) have remained uncleared. Much of the deforestation that has occurred in lowland areas has been for agriculture, especially sugar cane farms, which began in the late 1870s. In the uplands, clearing has been for cattle pastures, mainly on the Atherton Tableland. This too began in the nineteenth century, but had almost ceased by the 1920s. In both cases rain forest on fertile, basaltic soils was favoured for clearing.

Fire affects the Australian forests, especially at their margins. In the past, Aborigines used fire to manipulate the environment, changing rain forest into eucalypt forest and grassland. However, it is now possible to find areas where old eucalypt forest has been recolonized by rain forest. More recently, sugar cane farmers used fire to clear their fields. Uncontrolled burning of this sort has accounted for some of the deforestation along the edges of the forests. Once cleared, most of this land has been colonized by other types of forest.

The politics of protection

Until recently most of the rain forests between Townsville and Cooktown were managed by the State of Queensland as state forests or national parks. In the past, this protection prevented indiscriminate felling for agriculture. Some of these forests were selectively logged by the State on a 30–40 year logging cycle to provide a sustained yield of timber. The State's Forest Department claimed that this exploitation did no long-term damage to the ecosystem, and actually increased species diversity, a view that was hotly contested by conservationists. In the late 1940s it estimated the sustainable yield to be 75,000 cubic metres a year (2.6 million cubic feet). This rose to 600,000 cubic metres (21 million cubic feet) a year in the late 1960s, but was subsequently reduced to less than 80,000 cubic metres (2.8 million cubic feet) a year in the 1980s. These wide-ranging adjustments, while claimed to be based on sound advice and rigorously enforced, caused public unease. This, coupled with concern that the national park system was inadequate, led to moves by the federal government to have almost all of Queensland's rain forests listed as a World Heritage Site and to ban logging completely. The listing procedure was formally completed in 1988 amidst considerable controversy and strong opposition from the state government, which challenged the federal government's authority in court. However, the case was dismissed when the state elections at the end of 1989 brought a new administration into Queensland, sympathetic to the federal position. The only forests not to be included in the World Heritage Site proposal are the northern rain forests of the Cape York Peninsula.

Rain forest in Lamington National Park, Queensland.

The challenge of conservation

The human race has steered the planet Earth toward a serious state of imbalance, and no one can afford to ignore the fact that we are probably entering a period of unprecedented environmental crisis. Throughout the world, the very means by which all people can survive and prosper are being destroyed. In the developing countries, around a billion people exist in severe poverty that denies them their basic human needs. They look to nature to provide, but the burgeoning populations are too great for many tropical soils to support: reports of the resulting cycle of poverty, malnutrition and degradation of natural resources – which in turn increases poverty – are commonplace. The long list of disasters associated with the over-exploitation of tropical regions has become all too familiar – soil erosion, desertification, loss of cropland, deforestation, ecosystem destruction and extinction of species.

In the industrialized world, while populations are stabilizing, the consumption of fossil fuels, water and timber stocks, and other natural resources continues to grow steadily. Global energy consumption has increased by about 50 percent over the past two decades, almost entirely because of the ever higher demands of the industrialized world. In the mid-1980s the United States accounted for 25 percent of energy use, whereas sub-Saharan Africa took only one percent. The by-products of manufacturing industry are a legacy of pollution and climate change that will be inherited by future generations.

In the contest between people and the environment there are no winners. Our planet's capacity to support people is being irreversibly reduced at the very time when booming populations and growing consumption are making ever heavier demands on it. The root of the problem is people, who have always assumed dominance over nature: 7,000 years ago, the cradle of agriculture and civilization in the Middle East was deforested; the Greek and Roman empires severely degraded the Mediterranean regions; the Germanic, Celtic and Slavic tribes cleared most of Europe and then moved on to the New World.

Now the frontiers are in the tropics, notably in the tropical forests. Must the same patterns be repeated? Or can development strategies be found that reconcile the fulfilment of human needs with a respect for the natural world?

This is the challenge of conservation. It is far more than protecting individual species of plants and animals, more than keeping seas and freshwaters clean, more than caring for soils, more than protecting forest watersheds to ensure future water supplies, more than setting aside key sites as protected areas. Conservation is meeting all these objectives, but at the same time satisfying the basic human needs for housing, warmth, food, health and education. It is a form of management of natural resources to meet present human demand while maintaining sufficient stocks of renewable resources, and the ecological processes on which they depend.

As the twentieth century draws to a close we no longer have the excuse of ignorance that earlier civilizations had. The vulnerability of our fragile planet is public knowledge. We can save the world – if we have a mind to – but time is short. During the lifetime of many people living today, the world's human population has tripled. It will double again before today's schoolchildren finish their working lives. The earth *can* provide, but the laws of nature are strict and the penalties for breaking them catastrophic.

We can save the world – if we have a mind to – but time is short. The earth *can* provide, but the laws of nature are strict and the penalties for breaking them catastrophic.

Forest recovery – Rain forest that has been damaged by human activities (*main pic*) can, if left alone, recover . . . but few such areas once touched are left in peace. Surprisingly perhaps, the siting of an oil well in the forest (*inset*) causes little direct environmental damage. Oil revenue can, after all, be used to pay for conservation.

Global policies for global problems

The environmental challenges of coming decades are great indeed. But human powers of destruction and annihilation are matched by another great characteristic – the capacity to acquire knowledge and wisdom, and apply them creatively to solving the problems of life. Over the past two decades, and particularly in the 1980s, this creativity has been harnessed to develop conservation strategies for the future.

The United Nations Conference on the Human Environment that took place in Stockholm in 1972 was a historic meeting. For the first time the world's leaders acknowledged physical and biological resources to be a constraint on human development. The problems were spelled out in the *World Conservation Strategy* published in 1980 by The World Conservation Union (IUCN) with the support of the United Nations Environment Programme and the World Wide Fund for Nature (WWF). This document addressed government policy-makers, conservation groups and development practitioners from aid agencies, industry and commerce, and offered clear guidance on what ought to be done. It explained why conservation is crucial for human survival and sustainable development, identified the priority issues, and proposed effective ways of dealing with them (see box).

The *World Conservation Strategy* pinpointed the vital role that rain forests play. Their importance as a reservoir of biological diversity, coupled with the protective services that they provide – particularly their part in the maintenance of global climate – catapulted rain forests to the top of the list of world environmental issues. The *Strategy* suggests that they can be saved only after accepted ideas of conservation are re-evaluated. Conservation of living resources is often treated as a narrow, specialized activity, but in fact it is a process that cuts across all human activities, and must be planned as such. Conservation and human development need to be fully integrated to ensure that, in their quest for a better life, people *protect* those parts of the living world that are fragile, and *modify* the rest only in ways that can endure.

Our common future

The *World Conservation Strategy* set out to change the way governments were thinking and acting. This process took a great leap forward in 1983, when the General Assembly of the United Nations accepted the environmental challenge and called for an independent commission to draw up a "global agenda for change". The World Commission on Environment and Development was set up under the chairmanship of Gro Harlem Brundtland, then Prime Minister of Norway. It was charged with working out a strategy for achieving sustainable development for the year 2000 and beyond, improving international cooperation in reaching this goal, and developing an agenda for action at the political level. Since the Commission's report *Our Common Future* was published in 1987, the principle of sustainable development has entered the world's political awareness. This awareness was developed further in 1987 when the United Nations Environment Programme tabled its *Environmental Perspective to the Year 2000 and Beyond*, which strongly reinforced the factual basis and the forward-looking philosophy of *Our Common Future*.

Many of the political objectives set out in the above reports have been achieved, yet the world's rain forests continue to suffer clearance for inappropriate forms of agriculture, poorly managed logging, and other forms of abuse and degradation. Much more action is needed to put principles into practice. In 1989 IUCN published *From Strategy to Action*, a response to *Our Common Future* that develops three general principles that build on the solid foundations of sustainable development:

* Going beyond a sectoral approach to planning. Natural resources do not respect the boundaries of government departments: policies and projects that benefit one sector may be disastrous for others. Thus a hydroelectric dam may be valuable as a source of power for a capital city, but have a devastating effect on the lives of the people whose lands are submerged, and on the downstream forests that formerly took their nutrient supply from regular flooding. A new approach is required to coordinate the decision-makers in different sectors.

* Encouraging international cooperation. Nations are linked together in a complex web of investment, trade and communication, so the consequences of one country's internal policies are felt by its immediate neighbours and often by others across the globe. Greater international cooperation is essential if the problems facing the world community are to be solved.

* Building self-reliance. All too often, human societies have taken consumption and industrial production as the measures of social success. But the real objective must be quality of life, which encompasses health, security, literacy, longevity, and particularly the state of the environment in which people live. Future paths of development need to be mapped out in consultation with those who are directly affected. Local responsibility for local resources leaves room for adaptability to change, and promotes community involvement in questions of resource management.

Many of the world's successful development programmes have acknowledged and applied these principles; while a long list of environmental disasters testifies to the consequences of ignoring them. The application of these principles to sustainable development in tropical rain forests needs to be further clarified and summarized in order to give a clear lead for the future. This is what the remaining pages of this book set out to do.

A matter of scale (*right*) – It is difficult for many urban Westerners to appreciate the scale both of the rain forests as they once were, and as they are now. Monitoring of forest areas often requires travel by air, from fuel-base to fuel-base.

The World Conservation Strategy
To MAINTAIN ESSENTIAL ECOLOGICAL PROCESSES AND LIFE-SUPPORT SYSTEMS (such as soil regeneration and nutrient recycling), on which human survival depends.
To PRESERVE GENETIC DIVERSITY (the range of genetic material found in the world's organisms), from which we derive great benefits (see pages 32–33).
To ENSURE THE SUSTAINABLE UTILIZATION OF SPECIES AND ECOSYSTEMS (notably fish and other wildlife, forests and grazing lands), which support millions of rural communities as well as major industries.

The role of nations
As a global framework, the *World Conservation Strategy* inevitably takes a generalized approach. Putting the Strategy into action requires a national perspective, for it is at this level that major planning decisions affecting the environment are made. To date, few governments have taken adequate account of conservation objectives when planning development, and fewer still regulate their living resources to ensure their best sustainable use. For the past decade IUCN has been helping governments to prepare National Conservation Strategies that cross the usual planning boundaries between agriculture, forestry and industry, integrating development objectives with land-use planning that respects the natural limitations of forests and other wildlands. National Conservation Strategies always call for biologically rich and ecologically fragile lands to be put under protection – but the underlying principle is that all lands should be used in ways that do not cause permanent degradation. This is the way to hold open our options for the future.

Planning to conserve

It is inevitable that the populations of tropical countries will continue to rise and that their rain forest carpets will be rolled back still further. But growing human activity need not be incompatible with retaining the natural richness of rain forests. Current ideas of conservation go well beyond mere hands-off preservation: careful land-use planning at a national level allows the sustainable extraction of natural resources while maintaining the forest's biodiversity with all its potential benefits for the future. In 1971 the United Nations Educational, Scientific and Cultural Organization (UNESCO) began to put these ideas into practice with its Man and Biosphere Programme (MAB) to promote the setting up of a worldwide system of "Biosphere Reserves". These reserves are intended to be representative of natural ecosystems, but particular emphasis is placed on harmoniously integrating traditional patterns of land-use by involving local people in decision-making. More than 110 countries are taking part in the MAB Programme and 276 Biosphere Reserves covering 1.5 million square kilometres have been declared in 71 nations; about one-fifth of these are in tropical rain forests. UNESCO's role is to promote and monitor the programme through national MAB committees, but these have no legal powers.

The model forest

Ideally, lands are to be earmarked for their most appropriate usage as determined by soil quality, topography, accessibility and biological value. The resulting patchwork of land-use zones represents the whole spectrum of degrees of human impact. Inviolate conservation areas, set aside in their pristine state as national parks with access strictly controlled, lie at one extreme. Forests managed sustainably for the production of timber, latex, fruits, rattans and medicinal plants come next and are best sited as belts surrounding the pristine core areas. Plantations of rubber and oil palm or trees grown for their fibre or cellulose can also be part of this belt. More intensive land use, such as permanent arable farming, pasture and shifting agriculture are further parts of the jigsaw. Wetlands may be partly used for fish ponds or paddy fields. Then come towns, mines, artificial lakes and all the trappings of civilization.

Conservation of the natural heritage occurs throughout a landscape like that described above. The national park cores contain undisturbed forest and are designed to preserve all its structural complexity, species richness and interlinkages: these rain forest areas also act as controls against which to monitor the long-term effects of manipulating other zones. Many features of the pristine cores are shared by managed forests, particularly where the two are adjacent. Some animals are able to survive equally well in both zones, so larger populations may be supported than in the core alone: in addition, they may act as seed carriers, constantly reintroducing depleted species into the managed zone.

Plantations, too, have their place in a conservation plan: we have seen that rain forests supply a number of "environmental services" – such as a continuous supply of clear water and the stabilization of soil on steep slopes. These are in fact provided by any continuous vegetation mantle, so plantations of perennial crops, if appropriately managed, can equally well fill these roles. It is only in the most intensively altered and inhabited zones that the natural heritage is largely absent, although even here species-rich orchards and home-gardens near villages are a feature of many parts of the tropics.

In reality, large-scale land-use plans are difficult to implement. Seldom does the planner start with a blank sheet – an uninhabited landscape without any occupants and their vested interests. Even in the state of Rondônia in Brazil, an agroecological zoning project in a virtually uninhabited region was overwhelmed by huge numbers of immigrants who surged up the newly-constructed highway from south Brazil. On the island of Palawan in the Philippines there is still hope for successful zoning. The mountainous island is heavily forested and a graded system of protective management has been proposed, from strict control in some ecologically critical areas to light control spread over the whole island. In this way the biological richness on the island of Palawan will be maintained, but at the same time tribal peoples will be catered for, watersheds protected, timber stocks assured and fisheries maintained.

One very important facet of successful planning is to respect the rights of forest-dwelling people. In the past, these have often been ignored by central governments or forest departments. Nomadic hunter-gatherers live in and are totally dependent on the rain forest, yet it has been common for governments to take the paternalistic view that these peoples should be settled in villages. Similarly, subsistence farmers on the forest fringe rely on the forest to provide them with meat, fruit, honey, poles, cordage and medicines. All too often planners in the distant capital have omitted forest peoples from their calculations and persuaded aid agencies to fund development programmes that make the same mistake. Just as an awareness has grown that natural forest has long-term value that can and should be conserved as a nation develops, so there is today an increasing awareness that remote rural tribes with traditional lifestyles should not simply be swept aside as though they did not exist.

Secondary growth – When rain forest is cleared for any reason, the new growth that eventually replaces it is rarely of exactly the same type as the original. Among the trees striving to dominate the water's edge here beside a tributary of the Amazon in Peru are some *Cecropia* trees with their thin, grey trunks.

NORTH
AMERICA

EURASIA

AFRICA

Equator

SCALE

0   1,000   2,000 miles

0   1,000   2,000   3,000 kms

• Rain Forest Biosphere Reserve

Former and Existing Tropical
Rain Forests of the World.

SOUTH
AMERICA

AUSTRALIA

Sierra del
Rosario
Sian Ka'an   Guanica
                  Luquillo
Montes Azules   Río
Cordillera Volcánica   Platano
Central   Sierra Nevada de Santa Marta
Amistad   Darien
Cinturon Andino
Noroeste   Yasuni
Huascaran   Manu
"Ulla Ulla"   Beni
        Pilon-Lajas

Ziama Bia
Monts Nimba   Omo   Dja
        Tai   Basse-Lobaye
Impassa-Makokou   Yangambi
        Luki   Odzala
              Dimonika   Volcans

Mananara
Nord

Dinghu

Puerto
Galera

Gunung
Leuser

Sinharaja

Siberut   Lore Lindu
        Tanjung Puting
Cibodas
(Gunung Gede-
Pangrango)

Biosphere reserves (*above*) –
These reserves have been
established all round the world in
an attempt to discover the
solutions to such problems as
excessive tropical deforestation,
atmospheric pollution, the
"greenhouse effect", and
desertification. Each reserve
contains at least one sample of
an ecosystem characteristic of
one of the world's natural
regions, in which its human
inhabitants represent an integral
component. The main purpose
behind each is to explore
methods of both conservation
and sustainable exploitation of
the resources it contains, thus
benefiting the indigenous human
population, the wildlife, and of
course the plant species in all
their diversity. As a model
ecosystem, the reserves provide
unique opportunities for research
and the training of scientists in
many fields of expertise, not least
that of land management. The
results of such measures should
be of inestimable use to future
generations around the world.

181

Natural forests for sustainable timber

Timber from tropical rain forests is used all over the world for many different purposes. The construction industry makes extensive use of hardwoods in structural elements such as piles, groynes and beams as well as interior features like floors, doors and window frames; and some decorative tropical woods are favoured by cabinet makers and manufacturers of the finest musical instruments.

Many tropical countries rely heavily on the revenue generated by timber exports. Production of this timber by the sustainable use of rain forests has considerable economic advantages over establishing timber plantations. Firstly, it is cheaper because labour and maintenance costs are lower. Secondly, the time of timber harvest is flexible, so the forester can wait until the market is right without incurring a financial penalty. In contrast, timber plantations require a large initial investment tied to a risky prediction of market needs many years hence. Thirdly, conservation of other forest goods and services can be achieved alongside timber production, so potential money-spinners such as medicines and fruit species are preserved for the future. But the most important argument is that biodiversity and habitats are retained in managed forest and not in plantations. It is not surprising that natural forest management for timber production is planned (if seldom yet achieved) throughout the world's equatorial rain forest belt.

Rain forests provide a continuous, sustainable source of timber only as long as their natural ecological limits – as dictated by the dynamics of forest regeneration – are respected (see page 66). The early exploitation of rain forests did not test these limits: only single trees, scattered through the forest were dragged out by man or beast. But today's machinery permits far greater rates of logging, which must be maintained to justify the high purchase price and running costs of the machinery.

In recent years the range of tree species considered to be of commercial value has increased, further stepping up the rate of forest exploitation. The result of these developments has been a profound change in the make-up of the forest: scattered small canopy gaps which allow the regeneration of slow-growing, shade-tolerant species have given way to more and larger gaps, which favour fast-growing, light-demanding species.

Forests throughout the world differ in their proportions of these two species groups, and thus in their capacity to regenerate following extensive logging. For example, the rain forests of the western end of the Malay archipelago (Malaysia, the Philippines, Brunei and western Indonesia) are dominated by one family of commercially valuable trees, the Dipterocarpaceae, many of which are light-demanding "big-gap" species. These dipterocarp rain forests are therefore very robust in the face of modern, highly mechanized logging because they can re-establish very rapidly after extensive clearance. Dipterocarps have captured a large proportion of the tropical timber market because these relatively fast-growing trees produce pale, soft, low-density wood that makes good veneer and plywood.

On the other hand, the West African rain forests contain numerous "small-gap" species – such as members of the Meliaceae family, the so-called West African mahoganies – although no one family is dominant here. Many of Amazonia's rain forests are similar to those of Africa: numerous shade-tolerant "small-gap" species predominate. If these forests are to regenerate their natural composition, only low-intensity logging is permissible. All these forests are vulnerable to mismanagement and greed. If

excessively large gaps are created in any rain forest, a tangled mass of woody climbers and commercially useless small trees such as *Macaranga* and *Cecropia* (see page 65) grow up.

Managing the forest

Tropical forests are managed under one of two basic types of system. Under the monocyclic (or uniform) system, all the larger trees are harvested in one operation. The non-commercial species are killed by poisoning and seedlings are relied on to grow the new forest. In a polycyclic (or selection) system, only some of the mature trees are removed and the half-grown adolescents left behind for the next cut. A monocyclic system causes greater damage and needs a longer interval between cuts – commonly 70 years in dipterocarp rain forests. It favours fast-growing big-gap species. Conversely a polycyclic system favours small-gap species since more of the original forest is left intact. Recuts are possible every 20–40 years, depending on how fast the retained adolescent trees grow.

The Malayan Uniform (monocyclic) System was once widely used in Southeast Asia and is still in use in Sabah, where almost pure stands of dipterocarps occur. Today, the Asian forests are mainly exploited polycyclically, as are all managed African and Latin American forests. This system appeals to the business community and government resource managers because it allows a higher frequency of recuts and also because the markets only want logs of large dimension, especially for peeling into veneers. It is also compatible with forest conservation and continued extraction of minor forest products, because the scale of disruption is limited and the full diversity of plant and animal species is retained. Such a system has been operating successfully in Myanma (formerly Burma) for more than a hundred years. As long ago as 1850 the teak forests were being carefully assessed for the development of working plans. The "Burma Selection System" requires felling of mature trees over a 25–40 year cycle with extraction by elephant. The forests of Pegu Yomas, north of Yangon (formerly Rangoon), are in their third or fourth cycle of extraction.

From the biological and technical points of view, the management of natural rain forests as a sustainable source of timber is entirely possible. Where problems arise, it is usually for political, social and economic reasons and not because the forests are too fragile to be manipulated. For example, a recent IUCN study in Indonesia showed that forest management plans are broadly sustainable, but that they are almost invariably breached when implemented. The reasons are complex, but include lack of long-term incentives to the concessionaire and lack of monitoring by the Forest Department to prevent unnecessary damage. To make such a system work, damage has to be strictly limited. The seedlings or adolescents which are to form the new forests must not be destroyed during logging; roads must be correctly sited and constructed to minimize the area they disturb and the erosion they cause. Recuts must not be allowed before there is adequate growth. Hunters and farmers must not be allowed to enter once logging roads have given easy access, otherwise animal populations will be decimated, or the forest destroyed for farmland. But almost invariably these cautions have not been heeded, resulting in much unnecessary damage. This has led to the view that rain forests are too fragile to withstand utilization and to impassioned pleas for embargoes on tropical timber. These protests need to be better focused to attempt to put right the underlying weaknesses in human societies which actually lead to the mayhem.

A trunk around a trunk (*far left*) – Elephants work on the teak plantation in Lampang, northern Thailand, their massive strength easily hefting the enormous trees wherever they are directed. The elephant baby follows its mother and so learns about its probable future role.

Dipterocarp trees (*left*) – These relatively fast-growing rain forest trees overhang the Tahan River in Malaysia. Characteristically, dipterocarps have leathery evergreen leaves and, when in season, clusters of colourful and fragrant flowers with twisted petals.

Plantations

Throughout the 1960s and early 1970s, forestry in the humid tropics went through a phase when plantations were seen as a panacea. Their potential to produce a uniform product, with high yields in small areas compared to natural forests, seemed very attractive. Plantations were established in many countries with a succession of "miracle" exotic species, including *Anthocephalus chinensis* and *Paraserianthes (Albizia) falcataria*. Similarly, subtropical species of pine from semi-arid Central America, mainly *Pinus caribaea* and *Pinus oocarpa*, were introduced to Africa, Madagascar, Malaysia and elsewhere. The plans of the foresters were soon confounded: hard experience proved that species from the strongly seasonal tropics fared very badly in non-seasonal humid climates – a biologist would say not surprisingly. During this phase perfectly good pristine or lightly logged rain forests were cleared at great expense and replaced by plantations. For example, on the island of Bali in Indonesia a large proportion of the island's rain forests were inadvisedly cleared for pine plantations.

Tropical plantation forestry is beset by many other problems. The extensive land clearance needed to set up a plantation is both expensive and difficult, and the newly-bared ground is prone to erosion and cannot hold mineral nutrients, which leach down out of reach of plant roots. Plantations are composed of pure stands of single species, making them extremely vulnerable to attack by pests and diseases, and invasion by weeds. Costly and time-consuming measures must be taken to maintain the plantation's productivity. Weed species such as wild banana in Malaysian pine plantations and *Cecropia* trees in Amazonian yemane (*Gmelina arborea*) plantations must be physically cleared on a regular basis, while shoot-borers, the larvae of the moth *Hypsipyla grandella*, are the scourge of mahogany (*Swietenia macrophylla*) plantations in Central and South America.

In common with other crops, tree plantations have a high demand for mineral nutrients from the soil. Some timber plantations in the humid tropics have already experienced nutrient limitation; artificial fertilizers may be needed if the land is to produce even one crop of timber. This is unlikely to be feasible and, in any case, fertilizers are primarily intended for use on food crops, not timber trees.

A place for plantations

It has recently been realized that the replacement of natural forest with plantations is a misguided policy from the point of view of both conservation and economics. Nevertheless, plantations do have their place: they can usefully be established on the scrublands and wastelands which disfigure most tropical countries. In these areas they can restore forest cover and nutrient cycles, which are the first steps in restoring rain forest tree species to sites from which they have disappeared. Once the trees mature, birds and arboreal mammals move in bringing the seeds of native rain forest species with them, which can then become established in the understorey. If the plantation is then managed for conservation rather than profit, and the invading seedlings are not removed, a new rain forest may, in time, develop. This is happening in Sri Lanka, where pines are used to restore degraded land.

Plantations can also be valuable as physical barriers and buffer zones around national parks, where they provide extra living space for forest animals, and a source of commodities for local peoples. They can be justified economically where they yield different products from the natural forest, or bring degraded or damaged land back into use. Sabah, and more recently Indonesia, plan to establish plantations of *Acacia mangium*, *Albizia* and perhaps some dipterocarps on part of the 40,000 square kilometres (15,400 square miles) of Borneo destroyed by an 18-month drought and huge forest fire in 1983.

Trees for plantations

The trees most suitable for plantations are those which naturally colonize any large gaps that appear in the forest canopy (see pages 64–65, 184–185). Many are natural pioneers, which establish and grow rapidly out in the open, where they are exposed to direct sunlight, high temperatures and dry conditions. They produce copious, easily-dispersed seeds, which can persist for many years in the soil, waiting for the right germination conditions. These species colonize landslips, river banks and other bare land, and have the great advantage that they occur naturally in pure stands, so they have some inherent resistance to disease and pests. Unfortunately, many fast growers also have broad, spreading crowns, and if planted too close together for natural crown development, their growth slows dramatically. The financial return from a timber plantation depends on the quantity that can be produced per hectare, so the closer the trees can be spaced the better. Therefore, from a commercial point of view, the best plantation trees are those which have narrow crowns: this group includes pines which have a narrow, single-stemmed – or monopodial – crown.

The rain forest matrix itself contains a number of fast-growing species, but they do not all have the same set of useful features as pioneers; for example, seeds may be produced infrequently, and be large and difficult to store. Most of these species do not occur naturally in pure stands, and most develop multi-limbed, broad crowns (known as sympodial crowns). Fast-growing dipterocarps are an important example of this group, and have therefore very limited potential as plantation species. There will always be a market for dipterocarp timber, but it is currently obtainable only from managed natural forests.

The most widely planted broadleaved species in the humid tropics are yemane (*Gmelina arborea*) and *Acacia mangium*. Yemane is related to teak (*Tectona grandis*) and native to the same region, continental Southeast Asia. It produces a pale, fine-grained timber suitable for furniture and for fibre, but needs a fairly fertile soil. *Acacia mangium* is a pioneer native to the Queensland forests. It grows very fast to produce timber suitable for pulping and has the merit that it establishes easily on very degraded sites. A few species of eucalypt (*Eucalyptus deglupta, E. grandis, E. urophylla* and hybrids of the last two) have been fairly widely planted but are susceptible to shoot-boring moths like those that attack mahogany, destroying the growing point of the tree. Obeché (*Triplochiton scleroxylon*), a West African pioneer, has promise, but has not yet been very widely planted.

None of these species produces a particularly attractive timber, and indeed very few fine cabinet wood species have yet been grown successfully in plantations. One notable exception is mahogany, which is widely grown in Fiji, Sri Lanka and Sabah despite being commonly afflicted by a shoot-borer and pinhole-wood-borer beetle.

Plantation seed-bed – Seeds of fast-growing *Albizia* trees are sought in gallery forest like this in the Bulolo Gorge and elsewhere in Papua New Guinea for use in plantations. The species is particularly quick to restore forest cover in areas that have undergone widespread deforestation.

184

Resources for our future

Rain forests are not only exploited for their timber. A multitude of so-called minor forest products still forms the basis of local rural economies, and until a few decades ago these were more in demand than timber in the international market. In 1938, timber and these other products were of roughly equal value in the Indonesian economy: now timber exports provide 95 percent of the revenue from that nation's rain forests.

Recent research in the Amazon Basin suggests that this trend could be reversed, that non-timber products could provide better profits than timber in the medium-term. The conclusions, described on page 30, need to be tested in a variety of different locations before there can be any certainty that this principle would hold true over large areas of forest, much of it relatively far away from local markets. Most foresters and economists are sceptical, believing that timber will remain the main money-earning commodity into the foreseeable future, and that management of timber stocks for sustained yield is the main challenge.

Nevertheless, non-timber products are of undoubted value, both internationally (rattans, gums, latexes) and on a local scale (medicines, fruits). The need for conservation of this cornucopia of commodities is an essential weapon in the armoury of the conservationist because of the millions of rural people who are known to benefit directly from their use. Most minor forest products do not enter international markets. Many get no further than villages on the forest fringe, where they play an essential life-supporting role. If this is to be maintained, national and international development agencies must implement forest management plans – such as selective polycyclic felling (see page 182) – that encourage the sustainable production of other products in addition to timber.

Nature's medicine chest

In 1988 an international meeting of more than 50 pharmacologists, economists and conservation biologists met in Thailand under the auspices of the World Health Organization (WHO), IUCN and the World Wide Fund for Nature (WWF) to develop guidelines on how to conserve and use medicinal plants. The "Chiang Mai Declaration" called for greater effort to catalogue and conserve medicinal plants and launched a programme to "Save the Plants that Save Lives".

A few major world drugs are derived from rain forest plants. The anti-malarial agent quinine is extracted from the bark of several species of the Andean tree *Cinchona*. African and Asian species of the shrub *Rauvolfia* provide reserpine, which is used to reduce high blood pressure and to treat mental illness. Several legumes, especially the Moreton Bay chestnut (*Castanospermum australe*) of Australia, provide castanospermin, which is showing promise as a drug to combat AIDS. Forest people are the guardians of a huge natural pharmacy and there are high hopes that, with their assistance, new drug plants will be discovered. It has been estimated that three billion people – 60 percent of the world's population – depend upon traditional medicines for their principle source of cures for illness. Most of these are plants; in India and China 80–90 percent of traditional medicines are plant-based, and Chinese herbal treatments alone employ 5,000 species. Throughout the world the forests are the richest source of medicinal plants. For example, in Kenya 40 percent of herbal medicines come from native forest trees. In Amazonia an ethnobotanical team has catalogued more than 1,000 species of plants used by Indians, many of them as medicines.

Lost in the smoke of time?

An additional problem that must be faced by development planners is that of cultural change: as people move away from traditional lifestyles, their knowledge of forest plants and animals is soon lost. Efforts have been made to record this enormous reservoir of information: in the New World, organizations like WWF, the National Geographic Society, the Royal Botanic Gardens at Kew in England, and the New York and Missouri Botanic Gardens in the United States are working to document ethnobotanical lore. A single tribe of Amazonian Indians makes use of more than 100 different species of plants for medicinal purposes, yet virtually none of these has been properly analysed. More than 90 different Amazonian tribes are thought to have disappeared since the turn of the century, mainly through low resistance to new disease and over-exploitation by colonists. Fast-disappearing species are a problem, but knowledge of how species may be used is disappearing faster still.

Conservation of internationally valuable forest products

The single most important non-timber product to be traded internationally is rattan – the stems of climbing palms from the rain forests of the Asia-Pacific region. About 150,000 tonnes (130,000 US tons) are traded annually, representing a value of approximately US$2,250 million. Rattans grow wild in the forest, clawing their way up into the canopy using ferociously spiny leaves and tendrils. They are collected by forest people and sold to traders, often ending up in the traditional centres of the rattan trade – Ujung Pandang on Sulawesi, and Singapore. Until recently, much of the unprocessed rattan was exported to Hong Kong and the industrialized world as raw material for the manufacture of rope, baskets and furniture; but since the mid-1980s the trade in raw rattan has been progressively restricted and manufacturing now takes place locally. Rattan exploitation has increased as roads have improved access to the forest, social controls have broken down and timber workers have sought to supplement their income. As a result, this forest resource is disappearing so rapidly that plantations have begun to spring up to satisfy the demand – for example, more than 40 square kilometres (15 square miles) in Sabah over the past few years.

Despite the commercial importance of wild rattans, there remains much to be learned. In the Gunung Mulu National Park in Sarawak, the local Penan people, who use rattans for weaving, building and cordage, have pinpointed several new species for scientists. This emphasizes the importance of keeping wild stocks of forest products in protected areas while maintaining the local knowledge too.

The forest's store

Many other commercially significant crops originated in rain forests, including bananas, citrus fruits, mango and sugar cane; and many familiar spices such as cinnamon, ginger and nutmeg were first found in the rain forests of the East. Coffee (*Coffea* spp.) originally came from the montane forests of Ethiopia, oil palm (*Elaeis guineensis*) from lowland Africa and cocoa (*Theobroma cacao*) from the western Amazon. In the past, the exploitation of many tropical fruit species was limited by the perishability of their seed: trees such as durian (*Durio zibethinus*), rambutan (*Nephelium lappaceum*) and mangosteen (*Garcinia mangostana*) in the east, peach palm (*Bactris gasipaes*) and the guarana vine (*Paullinia cupana*) in the New World were once of only local significance. All of these have long been grown in gardens and small plantations; indeed the durian and mangosteen have never been found growing in the wild. Now that modern communications allow rapid seed transport, the full potential of these species can be realized.

Resins and latexes abound in rain forests, but their commercial importance has declined since the advent of synthetic equivalents. Some are still of commercial value: manila copal from the *Agathis* conifers of the Eastern rain forests is used in specialized varnishes, and certain damars (resins) from dipterocarps of the same region are used in perfumes. Gutta percha (the latex of *Palaquium gutta*) was one of the first products obtained from the rain forests of colonial Singapore and Malaya: once important as an insulator in submarine cables, it still has a minor specialized use as a temporary dental filling material for which its curious property of swelling as it sets makes it suitable. The Amazonian species *Hevea brasiliensis* provides the most important and famous latex, para rubber, whose story is well known. Following the discovery of the rubber vulcanization process, the high demand for pneumatic tyres which accompanied the advent of automobiles led to the Amazonian rubber boom of the late nineteenth century. Huge fortunes were made from the tapping of wild trees throughout the Amazon Basin. The merchants spent their riches on fine mansions and even opera houses in Manaus and Belém, while the tappers themselves lived in conditions that were approaching virtual slavery. *Hevea* seed was taken to the Royal Botanic Gardens at Kew in London, and seedlings were sent from there to Singapore where growers on the failing coffee plantations were persuaded to cultivate this new crop. Malaysia became prosperous and plantation-grown rubber almost eclipsed the native Amazonian product. However, the descendants of the original Brazilian tappers continued in their trade, largely forgotten by the outside world until the recent onslaught on the Amazonian forests by cattle ranchers led to violent confrontations (see page 124).

As the earth's supply of fossil fuels dwindles, we will increasingly have to turn to the plant kingdom as a source of the complex organic molecules needed by the chemical industry. The largest and most extensive collection of these compounds is found in the world's rain forest regions: it is essential for the future that these forests be safeguarded.

Rattan, Singapore (*left*) – The useful applications of the rattan vine are of extraordinary number and diversity, which is why its commercial value has soared over the last two decades. The increase in value, however, has in turn led to an increase in exploitation, and finally to the need for conservation measures.

Traditional methods (*above*) – The milky white latex, tapped from the rubber trees, is traditionally coagulated into solid form by smoking over a fire, building up a congealed mass layer by layer by pouring the latex over a spindle. The rubber obtained from most commercial plantations is now coagulated using chemicals.

Protecting the forests

The first ever national park, Yellowstone in the United States, was established in 1872. Since then more than 5,400 protected areas have been established worldwide, covering more than 5.5 million square kilometres (2.1 million square miles). Most governments have now recognized the value of protected areas to their people, and in the developing world of the tropics there are already almost 2,000 such sites covering more than 2 million square kilometres (more than 800,000 square miles). Although no precise data exist, perhaps one-third of these sites include some tropical rain or monsoon forest within their boundaries.

Has enough been done to ensure the long-term survival of the world's biological diversity within these protected areas? The answer is a very definite No. The problems are three-fold: some countries still have no protected areas and no legislation to put them in place; many countries have too small an area under protection; and most countries in the tropics have insufficient resources to manage their protected areas properly.

Key rain forest countries still lacking protected areas of any kind include the Comoro Islands, Equatorial Guinea, Laos and the Solomon Islands. Countries with too small an area under protection include Myanma (formerly Burma), Papua New Guinea and the Philippines in Asia; Guinea-Bissau, Liberia and Madagascar in Africa; Jamaica in the Caribbean; Brazil and Guyana in South America. All these protect considerably less than ten percent of their lands, a figure widely accepted as the minimum needed to maintain national biodiversity.

Management of species-rich sites is crucial to their long-term maintenance, but virtually all rain forest countries are unable to devote sufficient resources to training and employment of national park personnel. There are many different ways to manage protected areas. IUCN recognizes eight categories, ranging from strict nature reserves where people are only allowed in for scientific research, to multiple-use reserves where controlled extraction of game, timber and other forests products is permitted. Such a range of management options within a national protected area system is useful since it allows flexibility to provide for the needs of local people living around and occasionally inside the park. Trained managers are essential if the integrity of protected areas is to be assured.

The long-term effectiveness of these reserves depends not only on careful management, but also on their location. It is clearly desirable for the entire range of rain forest types to be represented in conservation areas, but in many cases considerations of *realpolitik* have determined that reserves be sited where there is no competition for land use. Some, however, have been consciously chosen to conserve particularly diverse landscapes and rich ecosystems. For example, the Manu National Park in Peru which lies on the east flank of the Andes, spans the range from montane to lowland rain forest. In the mid-1970s, a proposal was put forward for a comprehensive conservation network throughout the Brazilian part of the Amazon Basin which would include known centres of species richness and endemism, the so-called Pleistocene refugia (see page 54). Similarly, the Korup National Park in Cameroon is in one of Africa's Pleistocene refugia and is exceptionally rich in plant species (see page 138).

Principles for conservation

The size and habitat diversity of a conservation area need to be determined according to sound biological principles. If a reserve can support only a small population of a species, then that population is susceptible to chance extinction through disease or local catastrophe. A large population spread over a diverse landscape contains greater genetic diversity, which can prevent the accumulation of harmful genetic mutations and safeguard the population's long-term viability. The problems of reserve size are magnified for those animals which have very large territories. This group includes birds such as hornbills (*Rhyticeros* spp.) and predators at the top of the food chain, such as the tiger (*Panthera tigris*). It has been realized that conservation of these species can occur not only in national parks, but also in adjacent buffer zones in which limited extraction of timber and other forest products is permissible. Even relatively intensive timber extraction of up to 100 cubic metres per hectare (1,470 cubic feet per acre), which disrupts the structure of the forest for many years, does not necessarily eliminate forest vertebrates, including those which live in the canopy. This realization greatly increases the potential for animal conservation.

Although there are success stories amongst the world's national parks, much remains to be done. Those reserves that exist only on paper must be turned into reality, and those already in operation must be protected from damage by local people, ambitious logging companies or plantation owners. Biologists must continue their investigations both of forests which are being used to supply useful products and of natural forests, producing detailed species inventories and monitoring population. In the future, individuals of rare species might be moved between reserves to maintain genetic diversity, and introduced to sites if their known habitats were doomed. Governments should be encouraged to make land-use plans that maximize the opportunities for species conservation by zoning tree plantations and production forests around national parks. Ideally, parks should be linked by forest corridors, such as those planned for Sri Lanka as part of its forest development strategy.

Training and education are important adjuncts to the establishment of national parks, for without staff to manage them and without the support of the local population, they are doomed to fail in the long term. It is also important to involve local people in conservation by maintaining traditional rights when a national park is created: for example, in Mulu National Park in Sarawak, hunting access for the local people has been guaranteed by the Forest Department. Similarly, alternative sources of forest products, such as buffer-zone forests or plantations, should be made available. Many rain forest areas could be developed for tourism, thus giving a boost to the national economy as well as creating jobs for local people, whose traditional jungle lore could be adapted to this new use.

Too often in the past, international aid to tropical countries has caused deforestation as a by-product. Following pressure from conservationists, the environmental damage caused by projects sponsored by the World Bank, the EEC and others is starting to penetrate the world's political conscience. It is hoped that present trends will continue and that aid packages will increasingly include provision for the conservation of the natural world in the humid tropics.

There is now an abundance of data on what parks need to go where. All that is lacking is the political will to put them in place. The coming decade is crucial: all the key protected areas must be in place by the end of the century. After that the spread of people into rain forest lands will have closed the door on many options that exist today.

Tiger reserves – India set up 14 tiger reserves as a result of Project Tiger, established after a tiger census in 1972 found a total in India of fewer than 2,000 individuals. Today, some 24,700 sq km (9,550 sq miles) has been put aside for tiger conservation, and the number of tigers is again rising in consequence.

International cooperation

The reasons for rapid tropical deforestation are complex, but are all related to the way in which governments manage the growth and development of the human societies in their care. Development patterns in most countries have resulted in financial and material benefits for relatively few people, at the expense of an impoverished majority. The characteristic effects of these patterns include rapid population growth; extreme concentration of landholdings that leave millions in search of land; limited job opportunities; over-extended international loans; and inappropriate land-use decisions that are based on earning hard currency rather than directly benefitting local communities. Many development projects have caused deforestation, both directly and indirectly, with dire effects on communities who value forests. This book contains numerous examples of development projects that have run into unexpected problems, the majority of them collaborative ventures between national governments and the international development banks or aid agencies of the industrialized world. Yet there is still a strong basis for hope. Every failure, although costly in terms of environmental damage and human misery, is a lesson learned, while over the decades solutions have been found.

So far, implementation of successful solutions to deforestation and land misuse has taken place on a small scale. To achieve real success in reversing deforestation, concentrated effort on a broad front is needed, involving both public and private sectors, from heads of state and government ministries to local authorities and community groups.

Governments must take the lead in developing new policies and forging international links that can focus resources and expertise on the rain forest problem and bring about the policy changes needed to make sustainable forest use a practical possibility. Two major initiatives during the last decade have led to much greater collaboration between nations, and give hope for renewed optimism – the Tropical Forestry Action Plan and the International Tropical Timber Agreement.

The Tropical Forestry Action Plan

The development banks (World, Asian, African and Inter-American) and the overseas development agencies of the industrialized nations have been far from satisfied with the performance of their projects over the past two decades. Increasingly exposed by environmental "ginger" groups, the project treadmill of throwing good money after bad has become an embarrassment. Worst of all, in the early 1980s the failure of rain forest projects was leading to a decline in the level of international funding for the forest sector.

In 1985 the United Nations Food and Agriculture Organization (FAO) met with IUCN, the World Resources Institute, the World Bank and the United Nations Development Programme (UNDP) to develop a new strategy for forestry in the tropics, based on international cooperation and increased investment. The result was the Tropical Forestry Action Plan (TFAP).

The essence of the TFAP philosophy is that because poverty is the root cause of tropical deforestation, the richer nations should fund projects to alleviate that poverty and thus reduce deforestation. There is a good deal of healthy scepticism about the capability of development agencies to carry out projects that will genuinely discourage deforestation, since most projects in the past have achieved the opposite. However, it is important to remember that the TFAP is a strategy that is *on offer* to developing nations, not a master plan being inexorably implemented by the world's power brokers. The basic objective of the Plan is to restore, conserve and manage forests and forest lands in such a way that they sustainably benefit rural people, agriculture and the general economy of the countries concerned, and within this objective, it has great flexibility and adaptability. It aims to tackle five priority areas of concern (see opposite page), each a crucial part of the battle to use tropical forests sustainably and reduce deforestation.

Putting the plan into action

The first step in implementing the Plan was to inform all the tropical nations of its existence and invite them to participate. Since 1986, 73 countries have responded, representing 57 percent of the 129 potential participants and containing between them about 90 percent of the world's tropical forests. This remarkable response is a measure of the wide acceptability of the TFAP. The next step was to produce a review of tropical nations' forest resources and decide what sort of projects would be needed. In general, this review is carried out by a team of experts comprising representatives from the subject country, and foresters, economists and others from the nations interested in funding projects. As of 31 January 1990, 20 country reviews had been completed and a further 42 were ongoing.

Despite an appearance of progress, these reviews are in fact a long way from establishing, let alone completing, the desperately needed improvements in the management of rain forest lands. Only nine countries have so far gone right through to the stage of implementing projects, and none of these are completed.

Monitoring progress

The TFAP is in its infancy, yet the pace of deforestation is so high that there is an urgent need to monitor its success. The administration of the Plan is carried out by a small team based at FAO in Rome, supported by twice-yearly meetings of Forestry Advisors from participating nations. These groups have analysed the level of investment of overseas aid and technical assistance and found that between 1984 and 1988, spending in the forest sector rose by 80 percent. This is an impressive rise, but more is needed. The TFAP estimated that US$810 million a year was needed during 1987–91, but in 1988 the level of spending was still only US$576 million. In reality, there are difficulties in disbursing such huge sums in tropical countries. All too often there is insufficient expertise to manage projects in the field.

There are other bottlenecks slowing down the progress of TFAP. Despite the new approach to integrating conservation and development outlined on pages 178–179, and the explicit adoption of this approach in the TFAP, too many of the projects being designed are very traditional in their approach. They tend to compartmentalize logging, agriculture and conservation and keep them separate, but these are exactly the barriers that have led to disaster in the past. Only by integrating the planning process can the future of the forests be safeguarded. There is a great need for multiple-use management objectives, combining conservation of nature with timber extraction and harvesting of rattan, bamboo, fruits, nuts and other resources of the forest. There are few examples of such projects in operation at present, yet the future of rain forests may depend upon them.

The TFAP is by far the largest initiative ever focused on the conservation and sustainable management of tropical forests. Nations throughout the industrialized and developing world are taking part and the energy, resources and enthusiasm being channelled into the TFAP process represent a massive commitment. Yet the TFAP is only a framework for action. Its interpretation and its success lie in the hands of individual governments and their leaders.

By the year 2000 the TFAP will have either made significant steps towards saving the rain forests for future generations, or it will have fuelled a disaster of global proportions.

The International Tropical Timber Organization (ITTO)

On the international markets a number of global commodities are managed through commodity agreements, including cocoa, jute and coffee. These agreements are generally mediated by the United Nations Conference on Trade and Development (UNCTAD), and their objective is to stabilize markets and prevent over-production by establishing quotas based on market intelligence. In 1983 a new commodity agreement, the International Tropical Timber Agreement (ITTA) came into being, but this one made significant departures from its forebears. The ITTA is the first commodity agreement that deals with a naturally occurring resource. Unlike agricultural crops, the management of tropical forests for sustained production of timber is constrained by the productivity and ecology of the soils and species concerned and, with a few exceptions, is not manipulated by intensive management techniques, such as adding fertilizers, raising seedlings artificially or planting in monocultures.

The aim of the agreement is to establish a system of consultation and cooperation between timber consuming and producing nations. This is needed to ensure that management and exploitation of the commodity (that is, timber) is compatible with the maintenance of the ecological services of the forest, such as watershed protection and climate control. In other words, tropical forests need to be utilized for their timber, but at the same time conserved for their general benefits and to ensure that sufficient stocks are maintained for future generations to harvest. This is not a new concept. Even in colonial days, foresters were known as "conservators of the forests", but it is the first time an international commodity agreement has acknowledged conservation in its objectives.

It is for this reason that many governments have looked to the ITTA, and to the International Tropical Timber Organization (ITTO) that administers the agreement, for bold collaborative steps to control deforestation. The ITTO started life rather slowly, but the pace of its activities has increased in recent years and there have been some valuable advances.

The agreement was adopted in 1984 by 36 timber-producing and 33 timber-consuming nations. Adoption, however, does not constitute full membership of ITTO, and to date 18 producing and 22 consuming countries are full members. Nevertheless, their credentials are impressive. Between them the members control more than 80 percent of world Gross Domestic Product (GDP), and about 70 percent of the world's tropical forest resources. The consuming countries that are members account for more than 95 percent of the world's imports of tropical timber.

The objectives of ITTO are such that both timber trade and environmental organizations have supported the organization, attended meetings as observers and offered advice and information. All are interested in conserving resources for the future.

ITTO projects

The ITTO secretariat, based in Yokohama, Japan, has implemented a number of useful projects on behalf of the Council and committees of the organization. In collaboration with the Brazilian government, a preliminary project is being carried out in Acre to pave the way for a US$15 million proposal to develop forest industries in the state on an ecologically sustainable basis. The need for such projects, which can demonstrate how tropical forests should be managed, is very urgent indeed. Few of the major industrial logging enterprises currently operational are sustainable in the long term, as has been demonstrated by another ITTO project.

When an experienced team of foresters was despatched by ITTO to investigate current logging practices in the timber-producing member countries, they returned with depressing news. Of the roughly 8.3 million square kilometres (3.2 million square miles) of potentially productive rain and monsoon forests, the team concluded that less than 0.1 percent is currently being managed sustainably for timber (see box).

ITTO's future role

Although ITTO's projects have come up with valuable reports and initiatives the policy changes needed at national level, and the international cooperation and consensus needed to encourage these changes, have been slow to materialize. It is widely held that ITTO should be promoting the case for sustainable management of rain forests to member governments more strongly, demonstrating mechanisms for achieving this end, and setting achievable targets and timetables for implementation. Timber-producing countries are still competing directly with each other and thus encouraging the present low-priced buyers' market. Realistic timber pricing is essential if the full cost of sustainable utilization of the forests, including environmental costs, is to be met by those who use the timber.

There is no doubt that too much tropical hardwood is being used too quickly. Superb, close-grained timbers are being turned into disposable plywood used as the shuttering for pouring concrete structures, simply because they are so cheap. A reduction in output should accompany more realistic pricing, but such changes have to be made through international cooperation, in an equitable manner. ITTO is in a position to mediate such initiatives and should do so, soon.

The five priority areas of the Tropical Forestry Action Plan

FORESTRY IN LAND USE. Making better use of forest lands to ensure sustainable practices in forestry and agriculture, thus reducing soil erosion and increasing productivity.

DEVELOPMENT OF A SUSTAINABLE LOGGING INDUSTRY. Improving the ways in which tropical timbers are managed, including less damaging harvesting methods, reducing waste, and the development of more equitable and stable markets to meet both domestic and international needs.

MEETING FUELWOOD NEEDS. Resolution of the fuelwood crisis through a combination of conservation measures, encouraging use of alternative fuels and development of new fuelwood resources. (This problem applies more to seasonal forests and woodlands than to rain forests.)

CONSERVING FOREST SPECIES AND ECOSYSTEMS. Improving the protected areas systems in tropical forests in the interest of maintaining biodiversity, stabilizing local climate and hydrology, and providing the goods and services that human inhabitants of the forest need.

BUILDING KNOWLEDGE AND EXPERTISE IN FOREST MANAGEMENT. Providing support for forest research and forestry management institutions, with a view to improving standards and developing close cooperation between scientists, planners and decision-makers.

What are the problems?

The biggest problem is protecting the logged-over forest from incursion by agricultural colonists. Clearance after logging destroys any opportunities for sustainable management for timber. In addition, control of the logging operations represents a major difficulty. If carried out carefully, logging need cause little damage, but most concessionaires are poorly supervised and cause unnecessary damage to watersheds, soils and seedlings.

The market for timber is unbalanced. The present operation favours high volumes of low-value timber, when it should favour low volumes of high-value timber. The timber operation must also become profitable to those living in or near the forest, as well as to distant entrepreneurs.

To be solved, all these problems need up-to-date information on the status of the forest and the market. Programmes of survey, research and monitoring must be implemented to give the best possible chance of very much-needed success.

Think globally, act locally

"We call for a common endeavour and for new norms of behaviour at all levels and in the interests of all. The changes in attitude, in social values, and in aspirations will depend on vast campaigns of education, debate and public participation.

"To this end, we appeal to citizens' groups, to non-government organizations, to educational institutions, and to the scientific community. They have all played indispensable roles in the creation of public awareness and political change in the past. They will play a crucial part in putting the world on to sustainable development paths, in laying the groundwork for Our Common Future."*

Our Common Future, published by the World Commission of Environment and Development in 1987.

Governments usually have the final say on the fate of tropical forests, but for policies to be effective they need the support of the people at grass roots level. All too often there are conflicts of interest, and all over the world like-minded individuals have banded together to create non-governmental organizations (NGOs) able to put the case for the people. As the twentieth century draws to a close, NGOs number tens of thousands and represent untold millions of members and subscribers – a force to be reckoned with.

No one knows just how many NGOs have links with tropical forests worldwide, but a 1987 publication from the International Tree Project Clearinghouse lists more than 200 in Africa, and Friends of the Earth Malaysia has estimated about the same number of NGOs active in the Asia-Pacific region. They range in size from small, local pressure-groups anxious to maintain their immediate environment to international organizations such as the World Wide Fund for Nature (WWF), with national organizations in 25 countries on five continents.

Many NGOs act as pressure-groups, constantly researching for new information and lobbying the public and government for higher environmental standards and a more discerning approach to the exploitation of nature and natural resources. Foremost among these is Friends of the Earth (FOE), which maintains national organizations in 38 countries and an international secretariat in London. FOE has been carrying on its tropical rain forest campaign since 1985 and is perhaps best known for its exposure of the tropical timber trade. By analysing the trade statistics, FOE has drawn public attention to the main importers of tropical timber in Europe, North America and Japan. Today, people who buy tropical hardwood timber products in the temperate world are much more aware of their responsibilities and the issues behind them – largely through FOE's pioneering work in this area.

Other non-governmental organizations raise funds to put important projects into the field in rain forest countries. Often these are very specific, perhaps to protect an important forest or reserve, or to survey the status of threatened animals and plants. There are literally hundreds, possibly thousands, of NGOs operating in this way, but just a few examples will demonstrate the type of front-line work being undertaken.

*Wildlife Conservation International, a division of the New York Zoological Society, has projects throughout the tropics. Particularly noteworthy is a survey of the forest elephants (*Loxodonta africana cyclotis*) of west-central Africa, where populations are estimated to be 400,000 and declining. Similar work in Borneo has drawn attention to the declining range of the endemic proboscis monkey (*Nasalis larvatus*).

*The Fauna and Flora Preservation Society (FFPS) is one of the oldest NGOs. Founded at the turn of the century and based in the United Kingdom, it has an excellent track record in achieving conservation in the field. A long-standing commitment to gorilla conservation in Rwanda includes fundamental education work, as well as the maintenance of protected areas.

*The Nature Conservancy (TNC) has offices throughout the United States, where it has massive land management responsibilities and environmental databases. Over the past decade TNC has assisted local NGOs throughout Latin America in setting up national data centres and strengthening national conservation programmes.

*The Malayan Nature Society (MNS), founded in 1940, has pioneered nature conservation in Peninsular Malaysia. As well as spearheading education programmes and supporting existing reserves such as Taman Negara, MNS has taken a lead role in the effort to set up a new park in the Endau-Rompin area, one of the last strongholds for rhinoceros in the country (see page 161).

*The Nigerian Conservation Foundation (NCF) is becoming increasingly important in a country where rain forests are under severe pressures. The Foundation is making a determined effort to save the last unspoilt area, in the Oban hills in the south-east of the country. Their efforts were rewarded by the recent announcement of a new national park there (see page 137).

*The World Wide Fund for Nature (WWF) is undoubtedly the best-known international conservation NGO. Over the past 30 years WWF has spent well over US$100 million on more than 4,000 projects in 130 countries. Most of these projects have had national counterparts in the target country. WWF's work ranges from surveys identifying key areas to species protection, protected area establishment and management for conservation. Education of young people and training of personnel in nature conservation departments have also been high priorities. Over the years, WWF has worked closely with IUCN in establishing priorities and promoting the objectives of the *World Conservation Strategy* (see page 178).

Since 1986, tropical forest conservation has been one of WWF's three primary goals, but rain forests have been a priority since the Fund's inception. One of WWF's first projects, in 1962, involved the demarcation of a forest reserve in Madagascar. Today work continues in key rain forest areas such as Manu in Peru, which contains 10 percent of the world's bird species, and Korup in Cameroon, a rich forest area that survived the shrinking of the rain forests in the last Ice Age. Other projects include collaboration with Wildlife Conservation International in protection of the okapi (*Okapia johnstoni*) in Zaire; conservation of the Impenetrable (Bwindi) forest in Uganda, the Sapo forest in Liberia and the Täi forest in Côte d'Ivoire, the cross-border La Amistad national park in Panama and Costa Rica; and development of a management plan for Xishuangbanna Reserve, one of the last rain forests in China. WWF also helps to prepare conservation strategies, notably in the states of Malaysia, and, in the same region, has supported pioneering work on the impact of logging on wildlife.

*The World Conservation Union (IUCN) is unique in being both a non-governmental and an inter-governmental organization. As a union of governments and NGOs in 120 countries, its task is to promote conservation by building consensus of thought and action around the world (see back flap for details). IUCN has a Tropical Forest Programme, supported by Sweden, that concentrates on strategic planning of conservation, involving enough projects to show how strategies can be turned into reality. Field projects concentrate on reconciling conservation requirements with those of people living in forest areas, and special emphasis is given to the development of multiple-use buffer zones. This represents one of the great challenges to IUCN – demonstrating in the real world that the concepts and strategies developed in collaboration with advisors can be translated into genuine conservation action.

In coming decades the role of NGOs will be more and more crucial to the success of conservation and development projects. Governments and development banks increasingly need field intelligence to make sure that money is being well spent on the ground. Only when projects meet the needs of local people will the result be sustained in the long term. This applies as much to the establishment of a national park as it does to a new hydroelectric scheme. Over the past decade the world has seen NGOs spring out of nowhere to reach out to millions through the world's media. In some cases massive development programmes have been halted in their tracks. In the 1990s and beyond, both NGOs and governments must work together, building understanding from the early planning stage. In this way development money will be better spent, people in developing nations will be better served, and the rain forests will be better protected.

NGOs and the World Bank

The World Bank has for decades been a popular whipping-boy for NGOs, and justifiably so. The world's biggest spender in development assistance has an unhealthy reputation for being secretive and environmentally insensitive. For years projects with budgets of tens of millions of dollars were funded with only the minimum of environmental impact assessment. "Bank-bashing" by NGOs fed up with the waste, environmental destruction and all-too-often negligible benefits to rural people of the Bank-subsidized roads, mines, dams and cattle ranches, rose to a crescendo in the early 1980s.

Then in 1987, the World Bank set up its own Environment Department, pledging to work more closely with NGOs and take account of their grass-roots knowledge and community-level sensitivity. The Department has taken steps to find out where the world's richest and most threatened sites for biological diversity are located, sites that environmentalist Norman Myers dubbed the world's "hot-spots". In 1987 the World Bank convened a Task Force on Biological Diversity, with over 15 NGOs of international standing, including IUCN and WWF.

The World Bank pointed out that some US$50 million per year are potentially available for conservation from international donors, but that the bottlenecks in distributing these funds are in preparing good projects with integrated development and conservation objectives, and finding institutions strong enough to manage the projects.

NGOs can help with both of these aspects, and have been doing so in recent years. IUCN, WWF and others have been assisting in the preparation of project proposals and working in partnership with local institutions to put them into practice.

Madagascar, one of the world's most important "hot-spots", provides a good example of such cooperation. IUCN has been working with the World Bank to develop management programmes from the Andasibé and Ankarafantsika protected areas, and WWF has fielded a small team of experts to assist the government in identifying new sites and surveying the status of existing parks and reserves.

NGOs and the Tropical Forestry Action Plan (TFAP)

Meetings of the TFAP participants are open to observers from NGOs; IUCN, the International Institute for Environment and Development and the World Resources Institute have participated since the Plan's inception, with WWF joining more recently. An important contribution by these NGOs has been to prepare guidelines on the conduct of forest sector reviews, particularly in connection with the assessment of ecosystem conservation and the rights of forest-dwelling people.

Outside these meetings, however, relations with NGOs have been stormy. Many grass-roots organizations have seen the TFAP approach as too arrogantly patronising, both in its design and implementation. They feel that the architects of forest sector plans at the national level should take much greater heed of local communities and forest dwellers. Governments, through their policies, are forcing people into the forests to survive, then chastising them for what they do there. Yet they offer no realistic alternatives. Too many national Tropical Forestry Action Plans seem to do little more than fan the flames, NGOs say.

Major donors under the TFAP recognize the difficulties they face in spending money sensibly and to the best effect. Increasingly they are seeking partnerships with NGOs to assist them at the local level.

Swapping debts for nature

Until the early 1980s the developing countries enjoyed modest, but steady, growth in their economies. But since then the trend has been reversed. Worst hit was black Africa. In 1980 the average income per head was US$560 per head; in 1988 it was down to US$450. A major reason has been governmental mismanagement of national economies, and the biggest single area of misjudgment has been in international borrowing and lending.

Between 1980 and 1986 the total external debt of developing countries nearly doubled to over US$1 trillion. Although the rate of growth in debt has since slowed, the cost of repayments remains a massive burden, particularly for countries in Africa and Latin America. Recent World Bank statistics reveal how great the problem is: in 1989 the developing countries paid US$52 million *more* to the industrial world than they received in development assistance.

This anomaly is partly due to falling commodity prices: cocoa, rice, tea and sugar values peaked in the mid-1970s and have halved in value since. More importantly, the repayments on international debts are crippling. In Africa south of the Sahara, long-term debt was 58 percent of the region's GNP in 1986, and debt repayments took up 21 percent of the region's export income. Similarly in Latin America, the 1986 debt was 46 percent of its GNP, and debt service cost 30 percent of exports.

Among the hidden costs of the debt crisis are the environmental ones. As debtor nations suffer "structural readjustment" (for example, economic stringency measures) imposed by the International Monetary Fund, programmes for conservation of ecosystems and wildlife are among the first to go under. Throughout the world, many rain forest parks are protected only on paper; the money needed to manage them properly is simply not available. In 1984 in an article in the *New York Times*, Thomas E. Lovejoy formerly of WWF-US explained an idea that has helped to pull some parks back from the brink of obscurity.

While the world's bankers were struggling to reschedule debts and build allowances for bad debts into their annual reports, environmental NGOs were exploring ways of shouldering some of the debt burden themselves in return for concessions on national park infrastructure. This has proved attractive to some countries, for local currency can be used in the reserves, thus saving valuable foreign exchange. The first deal was struck in Bolivia in 1987 where Conservation International, an NGO based in Washington DC, acquired Bolivian debt of US$650,000, discounted to $100,000, from a Swiss bank. Conservation International then paid off the debt and in exchange the government of Bolivia set up a 14,000 square-kilometre (5,400 square-mile) buffer zone around the 1,340 square-kilometre (517 square-mile) Beni Biological Reserve. In Costa Rica and Ecuador WWF-US and The Nature Conservancy respectively acquired debts in return for long-term local currency bonds which they can use for conservation.

The crystal ball

People the world over *know* that if the rain forests are cleared humanity will lose untold benefits and suffer environmental degradation never before experienced. What are the achievable objectives and what new initiatives can be put in train?

* Move the planning of tropical forest use higher up the political agenda in tropical countries and ensure that a cross-sectoral approach is developed to integrate forestry with industry, agriculture, energy, immigration and other development plans.

* Set aside at least 15 percent of each country's *original* coverage of rain forest as totally protected areas, to ensure continuity of all forest ecosystems. In heavily populated areas with rich and fertile soils, lower targets of five or ten percent may be practical, but some other countries with large areas that can support only forest ecosystems could achieve 20 percent or more.

* Plan and manage the protected area system to conserve the populations of all rain forest species.

* Maintain a further 30–60 percent of larger nations' forests in a permanent, physically and legally secure forest estate. This will ensure supplies of timber for national development and of forest products for development at local levels, as well as securing water supplies to cities and agriculture, and allowing continuity in the traditional lifestyles of forest-dwelling people.

* Use plantation forests to produce much of the world's utility-grade timber and pulping wood, and to restore already degraded rain forest lands. The value of plantations as buffers around protected areas should be exploited more fully.

The two most important global initiatives to bring these objectives to reality are the Tropical Forestry Action Plan and the International Tropical Timber Agreement, both well under way (see page 191). Waiting in the wings are more programmes intended to complement and reinforce these existing endeavours.

The focus of the 1980 World Conservation Strategy (see page 178) assumed that reinforcing strategies on population, energy, food supply, economic development and human rights would also be prepared. Such support was not forthcoming and so the new Strategy for the 1990s, now being launched, takes a broader approach. It will tackle such issues as global climate warming, acid rain and the destruction of the ozone layer – all of which affect rain forests – in an ambitious, yet practical, way.

In 1982 the third World Congress on National Parks and Protected Areas in Bali, Indonesia, pledged to expand the global network of protected areas and to improve their management. In 1992 the fourth Congress will be convened in Venezuela. As the scope for new protected areas begins to shrink towards the end of this century, the 1992 Congress, destined to be the largest ever, will focus on resolving the conflicts inherent in conservation and the likely impact of global warming on the protected areas.

International conventions on wildlife trade, wetlands and World Heritage sites are able to boast of considerable success. Now a new Convention on Biological Diversity is being prepared by IUCN with the intention of laying down an obligation for all nations to ensure the conservation of their species and natural habitats, with financial assistance where necessary.

IUCN, the World Resources Institute and UNEP, in cooperation with more than 20 collaborating agencies, are developing a Global Biodiversity Strategy to tackle head-on the problem of disappearing species. It will identify ways to bring the benefits of conservation to local people, encourage better laws, and focus finance and international cooperation on the problem.

The culmination of these and many other initiatives to save the world's biological heritage will be the United Nations Environment Conference to be held in Brazil in 1992. Taking place 20 years after the Stockholm United Nations Conference that did so much to raise environmental awareness, it must now place environmental *management* at the forefront of political action. This will bear fruit only if it is supported at every level of society. Governments and conservation organizations cannot take unilateral action and succeed. They need help and collaboration from ordinary people around the world.

In central Brazil, the Xingú National Park in which these Indians live offers them only limited protection from outside interference.

Glossary

arboreal – of trees; in the trees; tree-like.

aroid – member of the plant family Araceae (the arum family).

biomass – the total dry weight of all living organisms in a specified area.

biosphere – the part of the earth's surface, including the oceans and the lower atmosphere, in which life exists.

biosphere reserve – conservation area in which an ecosystem characteristic to at least one specific region of the world's natural history is preserved for monitoring and scientific research.

blackwater river – river that is stained by rotting vegetation, and is thus black in colour.

bromeliad – member of an ancient family of trees, the Bromeliaceae; today's best known variant is the pineapple.

buttress root – root that grows from the trunk of a tree above the ground, which provides support and increases the area over which nutrients are absorbed from the soil.

caatinga – a type of rain forest (also referred to as a heath forest) which consists of stunted trees growing on nutrient-poor, sandy soil.

cauliferous – budding on the trunk, as opposed to putting out buds and flowers on branches and branchlets.

coevolution – process by which, over millions of years, two species gradually adapt to each other in a way that is advantageous (and sometimes eventually vital) to both.

conifer – tree of the order of gymnosperms, which bear cones to reproduce: typical of the order are firs, pines and yews; almost all are evergreen.

cycad – plant of the order of gymnosperms, similar to the conifers but resembling palms or ferns.

dipterocarp – tree of the family Dipterocarpaceae; many are natives of Southeast Asia, and have large, leathery leaves and brightly coloured flowers. Dipterocarps flower infrequently, maybe once in every five years.

ecosystem – organization of vegetation and wildlife within a specific climatic zone.

endemism – in this book, the property of being found only in a specified region and nowhere else in the world; the adjective, also given this specialized meaning in this book, is endemic.

epiphyte – plant that grows on another plant (or, rarely, an animal) without taking nourishment from its host, using it instead as the means to obtain nutrients from a position it could otherwise not reach.

greenhouse gas – gas in the earth's atmosphere known to contribute to global warming.

gymnosperm – member of an ancient, even primitive, group of plants of which the seeds are not enclosed in an ovary; the term literally means "naked seed"; major gymnosperm tree families include the conifers and the cycads.

heath forest – forest of relatively stunted growth due to the absence of essential nutrients in the soil, or to altitude.

hunter-gatherers – people who obtain their food, and much else of their livelihood, by hunting wildlife or gathering vegetable food from the natural environment.

igapó – of the Amazon Basin; a type of lowland rain forest that is periodically flooded by a blackwater river.

leach out – drain out as liquid percolates through.

mycorrhizal fungi – fungi that attach themselves to roots and then reach out for decomposing leaves on the forest floor, channelling nutrients from the decomposition back to the roots.

New World – in general terms, the Americas and Oceania.

Old World – in general terms, Eurasia (including the Indian subcontinent) and Africa (including Madagascar).

photosynthesis – chemical reaction by which plants combine carbon dioxide from the air with water from the soil to make energy-rich glucose, releasing excess oxygen into the atmosphere.

prop root – root put out from the trunk on or above ground level to provide support to the tree for vertical growth.

rattan – various species of climbing palm.

refugia – areas in which varieties of plant and animal life are particularly diverse and plentiful, a situation that is thought to result from a fortuitous escape by such areas from the effects of the last Ice Age.

root mat – spongy mass of entwined roots and rootlets (with other vegetable and fungal matter) that covers the floor of some rain forests.

saprophyte – plant that grows and feeds on rotting vegetable matter, generally on the forest floor.

saprotroph – organism (particularly an insect, but also a fungus or bacterium) that feeds on rotting vegetable matter.

shifting cultivation – system of agriculture that depends on farming an area of cleared forest for a short period, pehaps even a few years, until the soil is virtually unproductive, before moving on to clear another area of forest for the purpose.

soil nutrient – mineral substance found in the soil that is used by plants for growth: nitrogen, phosphorus, potassium, calcium and magnesium; rain forest soils are notably lacking in many of these.

stolon – also known as a runner; a stem that grows along the ground and produces roots and shoots from the nodes or tip.

symbiosis – relationship between organisms that contributes to the wellbeing of both and harms neither.

toxin – a natural poison.

transmigrant – agricultural worker (and his family) officially encouraged by the government to move into a forested area, clear it, and farm it.

várzea – of the Amazon Basin; a type of lowland rain forest that is periodically by a whitewater river.

whitewater river – river that contains considerable quantities of silt which may be deposited on a floodplain when in spate.

Index

Acknowledgments

This book had its origins in a collaborative venture designed by the International Union for Conservation of Nature (IUCN) and the World Conservation Monitoring Centre (WCMC) to map the world's rain forests. The mapping project was generously sponsored by British Petroleum (BP), and special thanks are due to Dr Eric Cowell, formerly of BP Group Environmental Services, who gave great enthusiasm and support to the work.

WCMC's databases form part of the Global Resources Information Database (GRID) set up by the United Nations Environment Programme's (UNEP) Global Environment Monitoring System (GEMS) for the compilation of environmental data. In 1988-89 a generous donation of computer equipment was made by IBM to GRID. WCMC benefited from this donation by receiving an IBM PS/2 Model 80 personal computer, which became the heart of the Centre's Geographic Information System (GIS). More recently, the Centre has received a generous donation of a graphics workstation from Tektronix, which will enable the future expansion of the rain forest database.

For software, our gratitude goes to the Environmental Systems Research Institute (ESRI) in California. ESRI donated the ARC/INFO software used on the GIS. Jack Dangermond, President of ESRI, has been particularly helpful and interested in WCMC's activities. The UK distributors of ARC/INFO, Doric Limited, have also been very helpful.

A compilation of data such as that presented here would have been impossible without the support and encouragement of colleagues in IUCN and WCMC; thanks go to all of them. In rain forest countries very many people in government departments and conservation organizations have generously provided maps, published and unpublished information, corrections and advice.

The UNEP/GEMS/GRID offices in Nairobi and Geneva require special acknowledgment. At the Geneva office, researchers Alan Cross, Risto Païvinen and Ron Witt have been experimenting in the analysis of weather-satellite data for assessing tropical forest cover. Very generously, we have been permitted to use their unpublished results for parts of West Africa and southern Amazonia.

Thanks to Tony Morrison for permission to use the quotation on page 46 which comes from *Margaret Mee In Search of Flowers of the Amazon Forests* published by Nonesuch Expeditions.

Thanks to Elizabeth Kemf for the photograph and feature on Vietnam (page 159) and Damien Lewis for the photographs and feature on the Kilum Project (page 31).

Illustrations by Linden Artists and Michael Woods.

Picture credits

Thanks to Gretta Stevens, Wendy Cooper, Terue Tateyama, Michèle DePraz and Rosabianca Ligios for help with the pictures.

1: Michael Fogden/Bruce Coleman; 2-3: Michael Fogden/Bruce Coleman; 4-5: Michael Fogden/Bruce Coleman; 6: Sue Cunningham; 8-9: Ken Preston Mafham/Premophotos; 10: Stephen Dalton/Oxford Scientific; 11: Loren McIntyre; 12-13: Tony Morrison/South American Pictures; 13: (inset) Diana Richards; 14: (inset left) Sue Cunningham, (inset middle) Michael Fogden/Bruce Coleman, (inset right) Michael Fogden/Bruce Coleman; 14-15: K. Wothe/Bruce Coleman; 16-17: Gerald Cubitt/Bruce Coleman; 18-19: Loren McIntyre; 20: Sue Cunningham; 20-21: Gary Retherford/Bruce Coleman; 21: Gerald Cubitt; 22: (left) John Mackinnon/Bruce Coleman, (right) Gerald Cubitt/Bruce Coleman; 23: C.B. and D.W. Frith/Bruce Coleman; 24-25: Gunter Ziesler/Bruce Coleman; 25: (inset left) Norman Owen Tomalin/Bruce Coleman, (inset right) Frieder Sauer/Bruce Coleman; 26-27: Robert Harding; 27: (inset left) Sue Cunningham, (inset right) J. Mackinnon/Bruce Coleman; 28-29: Robin Hanbury-Tenison/Robert Harding; 29: (inset) Peter Frey; 30: (top) L.C. Marigo/Bruce Coleman, (bottom) Jesco von Puttkamer/Hutchinson Library; 31: Stan Abbott/Deforestation Awareness; 32-33: Michael Fogden/Bruce Coleman; 33: (inset) Peter Addis; 34-35: Loren McIntyre; 35: Tony Morrison/South American Pictures; 36-37: Loren McIntyre; 38: Peter Frey; 39: (top) Peter Frey, (bottom) Steve McCurrey/Magnum; 40-41: Peter Frey; 41: (top) J.B. Ratcliffe/WWF, (bottom) L.C. Marigo/Bruce Coleman; 42-43: Peter Frey; 44-45: Peter Frey; 45: (top) Abril Bre/Gamma, (bottom) Sue Cunningham; 46-47: Loren McIntyre; 47: (top) Christian Zuber/Bruce Coleman, (middle) Sue Cunningham, (bottom) Michael Fogden/Bruce Coleman; 49: (top) L.C. Marigo/Bruce Coleman, (bottom) Ken Preston-Mafham/Premophotos; 50: (top) Michael Fogden/Bruce Coleman, (bottom) Brian Coates/Bruce Coleman; 51: Dieter and Mary Plage/Bruce Coleman; 52: Rod Williams; 54-55: Philip Sharpe/Oxford Scientific Films; 56-57: Michael Fogden/Bruce Coleman; 58: O. Langrand/Bruce Coleman; 59: Gerald Cubitt/Bruce Coleman; 60: Michael Fogden/Bruce Coleman; 61: Francis Halle; 62: (left) Michael Fogden/Bruce Coleman, (right) Mark Boulton/Bruce Coleman; 63: (top) Ghillean Prance, (bottom left) Brian Rogers/Biofotos, (bottom right) Carol Farneti/Oxford Scientific; 64-65: Stephen Dalton/Oxford Scientific; 65: (top) Heather Angel, (inset top) Ken Preston-Mafham/Premaphotos, (inset bottom) Deni Bown; 66: Deni Bown; 66-67: Brian Rogers/Biofotos; 67: Alain Compost/Bruce Coleman; 68-69: Sue Cunningham; 69: (left) Chris Leigh, (right) Royal Geographic Society; 70: J. Mackinnon/Bruce Coleman; 71: (left) G. Gottsberger, (top right) Stephen J. Krasemann/Bruce Coleman; 72-73: Ken Preston-Mafham/Premophotos; 73: (inset left) Mark Collins, (inset right) Ken Preston-Mafham/Premophotos; 74-75: Michael Fogden/Bruce Coleman; 76-77: Stephen Dalton/NHPA; 77: (inset left) Michael and Patricia Fogden, (inset right) Norman Owen Tomalin; 78: C.B. & D.W. Frith/Bruce Coleman; 78-79: Hans Rheinhard/Bruce Coleman; 79: Rod Williams/Bruce Coleman; 80-81: Michael Fogden/Bruce Coleman; 81: Stephen Dalton/NHPA; 82-83: David Houston/Bruce Coleman; 83: Alain Compost/Bruce Coleman; 84-85: Michael Fogden/Bruce Coleman; 85: (top) Michael Fogden/Bruce Coleman, (bottom) Gunter Ziesler/Bruce Coleman; 86: C. Houston/Bruce Coleman; 86-87: Ken Preston-Mafham/Premophotos; 87: (top) Gunter Ziesler/Bruce Coleman, (bottom right) Erwin and Peggy Bauer; 88-89: Alain Compost/Bruce Coleman; 89: (top) Jean-Paul Ferrro/Auscape, (bottom left) Adrian Warren/Ardea, (bottom right) Gunter Ziesler/Bruce Coleman; 90-91: Robert Harding; 92-93: Robin Hanbury-Tenison/Robert Harding; 93: Moser, Taylor/Hutchinson Library; 94-95: Robin Hanbury-Tenison/Robert Harding; 95: Moser, Taylor/Hutchinson Picture Library; 98: Andi Spicer; 99: Telegraph Colour Library; 100: (left) Michael Freeman/Bruce Coleman, (middle left) Bruce Coleman, (middle right) Gunter Ziesler/Bruce Coleman, Sue Cunningham; 105: (top) Steven Kaufman/Bruce Coleman, (bottom) M. Timothy O'Keefe/Bruce Coleman; 106-107: Michael Friedel/Rex Features; 108-109: Frederick Ayer/Science Photo Library; 112: Sue Cunningham; 113: Roger Jackman/Bruce Coleman; 117: Hutchinson Library; 119: L.C. Marigo/Bruce Coleman; 120-121: Ghillean Prance; 122: Sue Cunningham; 124-125: Christian Aid; 125: Loren McIntyre; 128: (left) Leimbach/Robert Harding, (right) Sue Cunningham; 129: Robin Hanbury-Tenison/Robert Harding; 131: Heather Angel; 132: (top) Ken Preston-Mafham/Premaphotos, (bottom) K;H;Redford; 135: L.C. Marigo/Bruce Coleman; 138: P. Parker/Hutchinson Library; 139: (left) R. Campbell/Bruce Coleman, (right) Jane Burton/Bruce Coleman; 142: Peter Davey/Bruce Coleman; 144: R. Campbell/Bruce Coleman; 147: Ken Preston-Mafham/Premaphotos; 148: Jennifer Fry/Bruce Coleman, (inset) O. Langrand/Bruce Coleman; 149: Rod Williams/Bruce Coleman; 151: Francisco Erize/Bruce Coleman; 152: (top) Dieter and Mary Plage/Bruce Coleman, (bottom) Gerald Cubitt/Bruce Coleman; 153: Paul Wilkinson; 154: Gunter Ziesler/Bruce Coleman; 157: WWF; 159: Elizabeth Kemf; 160: Rod Williams/Bruce Coleman; 162: Michael Freeman/Bruce Coleman; 164: (top) Gunter Ziesler/Bruce Coleman, (bottom) Heather Angel; 165: (left) Tim Porter/Camera Press-WWF, (right) Andi Spicer; 169: (top) Brian J. Coates/Bruce Coleman, (bottom) Alain Compost/Bruce Coleman; 173: (left) Christian Dodwell/Hutchinson Library, (middle) Michael MacIntyre/Hutchinson Library, (right) Michael MacIntyre/Hutchinson Library; 175: Jan Taylor/Bruce Coleman; 176-177: Peter Ward/Bruce Coleman; 177: (inset) Sue Cunningham; 179: Sue Cunningham; 180-181: Heather Angel; 182: J.A. McNeely/WWF; 183: Gerald Cubitt/Bruce Coleman; 185: Brian Rogers/Biofotos; 186-187: Duncan Poore/ICCE; 187: Peter Frey; 189: Rod Williams/Bruce Coleman; 194-195: Robert Harding.

Map Sources

General: – Campbell, D.G. & Hammond, H.D. (eds), *Floristic Inventory of Tropical Countries: The Status of Plant Systematics, Collections, and Vegetation, plus Recommendations for the Future*, The New York Botanical Garden, New York, 1989. **General: Africa** – UNESCO/AETFAT/UNSO *Vegetation map of Africa* (map), (compiled by F. White) Unesco, Paris, 1981. **General: Asia** – *1)* Whitmore, T.C., *A vegetation map of Malaysia* (map), *Journal of Biogeography* 11: 461-471, 1984. *2)* IUCN *Review of the Protected Areas Systemin the Indo-Malayan Realm*, (consultants J. MacKinnon & K. Mackinnon), International Union for Conservation of Nature and Natural Resources, Cambridge & Gland, 1986. **General: South America** – Whitmore, T.C. & Prance, G.T. (eds.), *Biogeography and Quaternary History in Tropical America*, Oxford Science Publication, 1987. **The Amazon Basin** – Malleux, J., *A Harmonized Forest Types Classification System for the Amazon Region*, (unpublished). **Australia** – *1)* Webb, L.J. & Tracey, J.G., *Australian rainforests: patterns and change, Ecological Biogeography of Australia*, W. Junk, The Hague, 1981. *2)* Department of Forestry, *North Queensland* (map), Edition 4, 1987. *3)* Department of Forestry, *Far North Queensland* (map), Edition 1, 1988. **Bangladesh** – World Bank, *Bangladesh General Vegetation* (prepared by the Resource Planning Unit, Agriculture and Rural Development Department), World Bank, Washington, 1981. **Belize** – Hartshorn, G. *et al*, *Belize. Country Environmental Profile. A Field Study.*, USAID contract, JRB Associates, Virginia, 1981. **Brasil** – IBGE/IBDF, *Mapa de VegetaÇao do Brasil* (map), 1988. **Brunei** – Anderson, J.A.R. & Marsden, D., *Brunei Forest Resources and Strategic Planning Study*, Unpublished report to the Government of Negara Brunei Darussalam, 1988. **Myanma (Burma)** – *1)* Britto, *National forest management and inventory Burma. Report on cartographic consultancy.*, Forest Department of Burma/fao, Rangoon, 1987. *2)* Champion, H.G., *A preliminary survey of the forest types of India and Burma*, 1936. **Burundi** – AID/MAB, *Draft Environmental Profile of Burundi*, 1981. **Cambodia** – Legris, P. & Blasco, F., *Carte Internationale du Tapis Végétal et des Conditions Ecologiques/Cambodge* (map), Institut Francais, Pondicherry, 1971. **Cameroon** – Letouzey, R., *Notice de la Carte Phytogeographique du Cameroun*, Institut de la Carte Internationale de Vegetation, Toulouse, 1985. **The Caribbean** – Beard, J.S., *The Natural Vegetation of the Windward and Leeward Islands*, Clarendon Press, Oxford, 1949. **Central America** – Leonard, H.J., *Natural Resources and Economic Development in Central America. A Regional Environmental Profile.*, International Institute for Environment and Development/Earthscan, Transaction Books, New Brunswick and Oxford,1987. **China** – Hon, H.Y. (Ed), *Vegetation Map of China* (map), Chinese Academy of Sciences, Beijing, 1979. **Colombia** – Instituto Geografico "Agustin Codazzi", *Mapa de Bosques. Republica de Colombia* (map), Ministerio de Hacienda, 1983. **Cuba** – *1)* Morellet, J., *Problemas forestales en Cuba, Instituto forestal Latino-Americano. De Investigacion y capacitacion. Boletin No. 32*, Agosto-Diciembre, Venezuela, 1970. *2)* Smith, E.E., *The forests of Cuba*, Maria Moors Cabot Foundation, 1954. *3)* FAO, *Informe al Gobierna de Cuba sobre politica forestal y su ejecucion*, CUB/FO FAO/58/8/6006, FAO, Rome,1958. **Dominican Republic** – *1)* FAO, *Republica Dominica. Inventaio y Fomento de los Recursos Forestales.*, FO:SF/DOMB. Informe Terminal, FAO, Rome, 1972. *2)* Hartshorn, G. *et al*, *The Dominican Republic. Country Environmental Profile. A Field Study.*, USAID contract, JRB Associates, Virginia, 1981. **East Africa** – Hawthorne, W.D., *East Africa Coastal Forests*, Botanical Values, Human Threats and Conservation Profiles (unpublished). **Grenada** – Weaver, P.L., *Forestry Development in Grenada. Technical Cooperation Programme. Technical Report: Forestry Planning and Management Activities.* FO: TCP/GRN/8851, FAO, Rome, 1989. **Guadeloupe** – *Vers un Amenagement de la Forêt Soumise de Guadeloupe. Direction des services pour la Guadeloupe*, Office National des Forêts, 1976. **Haiti** – Cobb, C.E., "Haiti: Against all odds", National Geographic 72: 645-670,1987. **India** – *1) National Forest Vegetation Map* (map), Forest Survey of India, 1986. *2)* Das Gupta (ed.), *Atlas of Forest Resources of India*, National Atlas Organization, Government of India, Calcutta, 1976. *3) Forest map of South India*, Karnataka and Kerala Forest Departments and French Institute, Pondicherry, 1986. **Indonesia** – RePPProT, *National Overview of the Regional Physical Planning Programme for Transmigration* (map), Overseas Development Natural Resources Institute (ODNRI), Chatham, 1990. **Jamaica** – *1) Jamaica Resource Assessment*, Ministry of Agriculture, Kingston, 1982. *2)* Eyre, A.L., *Jamaica: Test Case for Tropical Deforestation*, *Ambio* 16 338-343, 1987. *3)* Asprey, G.F. & Robbins, R.G., *The Vegetation of Jamaica, Ecological Monographs* 23, 1953. *4)* Clarke, C.G., *Jamaica in Maps*, Africana Publishing Company, New York, 1974. *5)* IIED, *Jamaica. Country Environmental Profile*, Ministry of Agriculture, Kingston, 1987. **Kenya** – Doute, R., Ochanda, N. & Epp, H., *A forest inventory of Kenya using remote sensing techniques*, Kenya Rangeland Ecological Monitoring Unit, Nairobi, 1981. **Laos** – Lao PDR Forest Department, *Forest Management Map, Lao PDR* (map), Forestry Department, Vientiane, 1987. **Lesser Antilles** – ECNAMP, *Survey of Conservation Priorities in the Lesser Antilles. Preliminary Data Atlas.*, Eastern Caribbean Natural Area Management Program, Caribbean Conservation Association, the University of Michigan and the United Nations Environment Programme, 1980. **Madagascar** – Green, G.M. & Sussman, R.W., "Deforestation history of the eastern rainforests of Madagascar with satellite images", *Science*, (in press). **Mexico** – *1)* Estrada, A. & Coates-Estrada, R., "Rain forest in Mexico: research and conservation at Los Tuxtlas", *Oryx* 17: 201-204, 1983. *2)* Gerez, P. & Villela, O.F., *Conservacion en Mexico: Sintesis Sobre Vertebrados Terrestres, Vegetacion y uso del Suelo.*, Instituto Nacional de Investigaciones Sobre Recursos Bioticos, 1988. **Panama** – Cobb, C.E., "Panama: Ever at the crossroads", *National Geographic* 169: 466-492, 1986. **Papua New Guinea** – Paijmans, K., *Vegetation map of Papua New Guinea* (map), *csiro Land Research Series* 35: 1-25, 1975. **Peninsula Malaysia** – *1)* Forest Department, *Peninsula Malaysia: The Forest Area* (hand-coloured map obtained from the Forest Department in Kuala Lumpur in 1989; simplified version of the published *Forest of Peninsular Malaysia* based on 1981-1982 data published in 1986). *2)* Wyatt-Smith, "A preliminary vegetation map of Malaya with descriptions of the vegetation types", *Journal of Tropical Geography* 18: 200-213, 1964. **The Philippines** – Forest Management Bureau, *Natural Forest Resources of the Philippines*, Department of Environment and Natural Resources, Manila, 1988. **Puerto Rico** – *1)* Birdsey, R.A. & Weaver, P.L., "Forest Area Trends in Puerto Rico". *Research Note SO-331*, United States Department of Agriculture, 1987. *2)* Englerth, G.H. & Wadsworth, F.H., *Effects of the 1956 Hurricane on Forests in Puerto Rico, Caribbean Forester* 20: 1-2, 1959. *3)* Little, E.L. Jr & Wadsworth, F.H., *Common trees of Puerto Rico and the Virgin Islands.*, Department of Agriculture, Washington D.C., 1964. *4)* Birdsey, R.A. & Wadsworth, F.H., "A new look at the forests of Puerto Rico", *Turrialba* 35: 11-17, 1985. **Rwanda** – *1)* IUCN, *Conservation et Amenagement des Forets Naturelles de la Crete Zaire-Nil au Rwanda: Rapport de Mission*, 1983. *2)* AID/MAB, *Draft Environmental Profile on Rwanda*, (unpublished). **Sabah** – Sabah Forest Department, *Sabah Malaysia. Natural Plantation Forests* (map), 1984. **Sarawak** – Sarawak Forest Department, *Forest Distribution and Land Use Map* (map), 1979. **South America** – *1)* Hueck, K. & Siebert, D., *Vegetationskarte von Sudamerikas* (map), Fischer, Stuttgart, 1972. *2)* UNESCO, *Vegetation map of South America*, Unesco, Paris, 1981. **Sri Lanka** – Survey Department of Sri Lanka, *Sri Lanka: Chena Cultivation in the Dry Zone and Dense Natural Forest* (map), 1988. **St Lucia** – Caribbean Conservation Association and Island Resources Foundation, *St Lucia Country Environmental Profiles for the Eastern Caribbean*, Draft, 1988. **Tanzania** – *1)* Lovett, J., *Development threats to the Eastern Arcs Forests of Tanzania*, wwf/iucn, 1985. *2)* Lovett, J.C., *An overview of the moist forests of Tanzania*, Tanzania National Scientific Research Council Monographs, 1986. **Thailand** – Royal Forest Department, *Forest Types map* (map), Royal Forest Department, Bangkok, 1985. **Trinidad** – Beard, J.S., *The natural vegetation of Trinidad*, Clarendon Press, Oxford, 1946. **Trinidad and Tobago** – Annual Report of the Forestry Division for the year 1972. **Uganda** – *1)* Struhsaker, T.T., *Forestry Issues and Conservation in Uganda, Biological Conservation* 39: 209-234, 1987. *2)* WWF, *Conservation of Tropical Forest Wildlife in Uganda. Annual Report March 1987. Project 3235*, 1987. **Venezuela** – *1)* Alarcon, C. & Huber, O., *Mapa de Vegetacion de Venezuela*, 1988. *2)* Ministerio del Ambiente y de los Recursos Naturales Renovables Instituto Forestal Latino Americo, *Vegetacion Mapa No.12.* (map), 1959. **Vietnam** – *Cac Loai Thuc Vat Bi de Doa Dien hinh va Môt Vung Tap Trung* (map), Results of a forest inventory in 1987. **West Africa** – Päivinen, R. & Witt, R., *The methodology development project for tropical forest cover assessment in West Africa*, uncp/gems/grid, (unpublished). **West Indies** – Watts, D., *The West Indies: Patterns of Development, Culture and Environmental Change since 1942*, Cambridge University Press, 1987.